S0-CKI-214

CONSTRUCTION SAFETY PRACTICES AND PRINCIPLES

Robert X. Peyton
Toni C. Rubio

VNR VAN NOSTRAND REINHOLD
New York

Library of Congress Catalog Number 90-23604
ISBN 0-442-23742-1

Printed in the United States of America

Van Nostrand Reinhold
115 Fifth Avenue
New York, New York 10003

Chapman and Hall
2–6 Boundary Row
London, SE1 8HN, England

Thomas Nelson Australia
102 Dodds Street
South Melbourne 3205
Victoria, Australia

Nelson Canada
1120 Birchmount Road
Scarborough, Ontario M1K 5G4, Canada

16 15 14 13 12 11 10 9 8 7 6 5 4 3 2 1

Library of Congress Cataloging-in-Publication Data

Peyton, Robert, X.
Construction safety practices and principles/by Robert X. Peyton, Toni C. Rubio.
p. cm.
Includes index.
ISBN 0-442-23742-1
1. Building—safety measures. I. Rubio. Toni C. II. Title
TH443.P48 1991
690′.22—dc20 90-23604
CIP

Contents

Preface

Many safety and health books are written from a theoretical or academic standpoint, providing an in-depth study of various aspects of occupational safety. While this approach is both helpful and necessary, it is not the path taken by this work.

Construction Safety Practices and Principles draws its strength from the lessons learned and knowledge gained in the authors' thirty-plus years of day-to-day management and site safety experience. It is a book written by field safety people for field safety people, and other people interested in Safety and Health.

It is, therefore, written from a management or employer's perspective, concentrating on what works on the job site, dealing with what is necessary and practical instead of what looks good on paper.

Construction Safety Practices and Principles provides you, the person, with Safety responsibility, with information in a format that you can easily use. This approach shows you how to do the job more effectively.

This book is a practical reference, providing valuable information for everyone from seasoned safety professionals to those people starting out in their first safety position.

So, whether you are an executive, a manager, or a line supervisor involved with safety, or a worker looking for effective safety information, this is the book for you!

Some of the areas explored in this book include:

· Written safety and health programs
· Crisis management in construction
· The hazard communications standard
· Dealing with OSHA
· Workplace substance abuse programs
· Effective safety training techniques

Like any reference book, this is a starting point, not an ending point. The authors will be happy to discuss any aspects of this book with you or answer your questions. Please write us at the following address: P. O. Box 491144, Ft. Lauderdale, FL 33349.

Robert X. Peyton
and Toni C. Rubio

Chapter 1

The Need for a More Proactive Construction Industry

INTRODUCTION

Chapter 1 expresses the authors' view that the construction industry needs a more unified approach toward handling the increase in standards and legislation aimed at regulation construction work sites in the name of safety.

The absence of a unified and active construction industry effort in these areas places the industry in the position of reacting to events instead of shaping them.

THE NEED FOR A MORE PROACTIVE CONSTRUCTION INDUSTRY

Construction is an industry under attack. The weapons used in that attack are increased governmental regulations, court action, and legislation. The end result is increased employer burdens and responsibilities, enacted under the guise of safety and health improvements. Safety reform is a "mom and apple pie" issue easily sold to the Congress and the public and is difficult to decry. However, what are labeled safety reforms may not improve job-site safety and will reduce contractors' abilities to compete in the construction market.

Industry employer associations and their interest groups oppose increased regulatory and economic burdens to a certain extent. They cannot, however, reach their goals unless their tactics change.

The Achilles' heel of the construction industry is a general lack of an organized resistance. The construction industry does not have a cooperative organized effort to oppose improper and ineffective government regulations; nor does it use its clout to enact better ones.

The strength of construction—its diversity—is also its weakness. Construction is an industry comprised predominantly of highly competitive small businesses with employers already swamped with the day-to-day effort of just doing business. This leaves little time to expend the resources or money needed to affect the standard setting or legislative process.

Construction's opposition in this attack is highly organized, falling into two basic areas—the Occupational Safety and Health Administration (OSHA) and organized labor.

OSHA and employers mostly have an adversarial relationship, with OSHA in the role of police officer rather than educator. Sadly, the United States is the only Western industrial power that relies predominantly on an enforcement rather than on an educational effort to achieve workplace safety.

Although enforcement is at times a necessary tool, it is often counterproductive to cooperation. OSHA, as the predominant safety and health authority in this country, should be the first place employers turn to for regulatory, technical, or compliance assistance. In most cases, unfortunately, it is the last place employers choose. Employers often adopt the attitude that OSHA is an entity that will not provide answers, only problems.

OSHA's 1970 mandate to ensure workplace safety and health cannot function without employer assistance. Safety on the job is still largely a matter of voluntary compliance on the part of employers. Yet, OSHA chooses to maintain an adversarial posture and role.

The Occupational Safety and Health Administration cannot possibly inspect all construction workplaces. It has neither the construction knowledgeable manpower nor the financial resources. OSHA needs that employer voluntary compliance, and yet, OSHA's efforts to recognize or encourage such compliance are minimal.

Yes, there are compliance-assistance programs within OSHA, but their impact upon construction employers is negligible. In fact, as of 1989, there were only two construction employers participating in the OSHA Voluntary Protection Programs. Although OSHA does engage in a few educational programs, that effort does not compare in scope or budget to OSHA's enforcement efforts.

OSHA's regulatory actions are not always practical or suitable for the construction industry. Two recent examples amply serve to illustrate this—the Hazard Communication Standard and the ANSI Lift Slab interpretation.

THE HAZARD COMMUNICATION STANDARD

Because of court actions against OSHA, a regulation entitled the Hazard Communication Standard was expanded to include construction. The premise of the standard is admirable—to inform employees of the dangers of the chemicals they may be exposed to in the workplace. The reality of the standard for construction is different. This standard will require the maintenance of volumes of papers on each construction site—paper that will be difficult to maintain and will be of minimal use.

These papers, called Material Safety Data Sheets (MSDS), are produced by chemical manufacturers to provide product information. MSDS's intended to educate workers are often too complex to understand. They are, however, highly technical and constantly expanding to provide increased liability protection for manufacturers. One construction trade association analyzed Material Safety Data Sheets for their reading-comprehension level. In an industry employing mostly high-school graduates, these MSDS's were written at a college or graduate-study level. How, then, will the needed information be understood by workers?

Additionally, this standard requires extensive amounts of employers' time for compliance, time that will not be available to use on other safety concerns. More importantly, this time and money spent addresses a problem that does not occur all that often on construction sites. Most chemical-related injuries recorded in the construction industry come from a handful of seriously hazardous materials, not from the multi-thousands of unproven potential and possible hazards covered under this standard. It is likely that this standard will have a minimal impact on reducing the chemical-related injuries in the construction industry. The Hazard Communication Standard denotes a disturbing trend in the regulatory process—the setting of standards through court actions.

This court intrusion into the OSHA standard setting process is growing. The unions began by suing OSHA in order to bypass the standard setting procedures and to force the enactment of regulations. The regulated industries countersued to modify those regulations and make them workable. In the end, the workers and employers lose due to delays in the enactment and improvement of job-site safety and the creation of unrealistic and ineffective regulations. Organized labor's use of the court process has been so effective that an advisory group to the government has recommended that OSHA consider the potential for union court action before issuing any standards.

THE AMERICAN NATIONAL STANDARDS INSTITUTE

The American National Standards Institute (ANSI) A.10 Committee develops voluntary consensus standards for the construction industry. Contractors used to be the driving force behind this standard setting process, fully participating in the development of advisory standards to improve the safety of construction operations.

Today's contractors' participation in ANSI is at an all-time low. Standards, therefore, are being developed without adequate contractor input. It is no surprise that without industry input, ANSI standards do not always reflect the best interests of the construction industry.

A blatant example is the ANSI Lift Slab Standard. Lift slab operations involve the casting of concrete floor slabs at ground level. Completed slabs are then lifted and secured in place, within the building's framework. Under the old Lift Slab Standard, while slabs are being put in place in one area, construction continues in other areas of the building.

In 1988, the ANSI A.10 Committee, in a drive headed by organized labor, issued an interpretation of the Lift Slab Standard, originally written in 1970. That interpretation prohibited ancillary construction work until all slab-lifting operations within the structure were completed. In effect, all other construction in the building is halted. Since this increases lift slab construction time, it makes this construction method more costly and less competitive.

OSHA then incorporated that A.10 interpretation directly into a temporary standard regulating lift slab operations. This action may drastically change the future impact of all ANSI standards.

A voluntary consensus standard (the A.10 Lift Slab Standard), originally developed by construction industry experts, was significantly changed by a reinterpretation of the standard's language without the industry's input and subsequent adoption as an OSHA regulation. OSHA is also considering the addition of similar ancillary activity bans in other concrete operation regulations.

Contractors in lift slab operations found themselves with a less competitive business because of an A.10 Committee action. Other contractors in tilt slab and cast or pour in place concrete operations may find their business also affected. Yet, these contractors had no part in the A.10 process and were not consulted about proposed changes to these regulations.

Employers enact safety programs because they are good business practices. Effective safety programs lower costs and increase job productivity, creating better profits. However, to be effective, safety regulations have to be realistic and applicable within the industry regulated. Without widespread industry input, they cannot be.

A PROACTIVE PUSH

What the construction industry needs is a two-part approach. First, an industry-wide organized proactive push to influence the regulatory and legislative processes is necessary. Second, more direct employer participation in the standard setting and planning process is also needed.

Individual employers need to realize the importance of their participation in setting regulations and drafting legislation.

It is imperative for business that construction employers become heavily involved in the regulatory legislative and ANSI arenas. As business people, employers simply cannot afford to shirk responsibility by "passing the buck." Employers take heed—without your active participation, your business will continue to be adversely affected.

The construction industry can no longer afford the patchwork-style and noncooperative approaches. It is time for the construction-industry employer groups to put aside their differences and the emphasis on their own concerns in order to work together. The construction industry needs to work toward one main goal—to defend itself with a coordinated and united effort, free from special-interest influence.

Construction can accomplish this goal through the creation of a nonprofit construction industry watchdog organization, supported by industry contributions, i.e., a single group representing a united construction industry. This nonpartisan organization could employ a small group of construction-industry experts and their support staffs, an organization with one goal—to represent the full interests of the construction industry.

This organization would have the express purpose of ensuring industry participation in all phases of the growing construction-regulatory forum. This organization could also work with and provide technical information to regulatory agencies and members of the legislature. With the strength of the half-million construction employers in this country, it would have an immensely strong economic and political voice, a voice no government entity or politician could ignore.

Employer associations will argue that they already fulfill that function, and to some extent they do. However, they represent the interests of only their limited memberships. As such, they generally serve to fragment cooperative industry efforts. Often, association internal politics and egos, rather than industry needs, dictate actions. Association staffs are often small, and multiple, vague job descriptions are common. Therefore, the work load on the individual staff members can be quite heavy. Workers often do not have the time or freedom to address the important issues. Policy is strictly dictated by the narrow concerns of the association.

We can learn from our opposition. The representatives of organized labor and the government are professional meeting attenders, paid to represent their groups' interest on those committees. They can devote their full time to that end, assisted by a support staff.

Construction representatives, either from individual companies, associations, or groups, are not free to devote exclusive time or effort to meetings. They must perform committee-type work in conjunction with their other duties. This places the construction industry at a disadvantage.

A proactive construction industry effort would place the emphasis on practical safety innovations—safety methods and developments tailored to this industry's needs, using proven and practical methods that result in industry-workable regulations.

A proactive push would allow the construction industry to develop and sponsor new OSHA regulations and safety legislation through a cooperative industry/government effort. Without that proactive effort, the construction industry will continue to find itself in the position of reacting to events rather than acting to initiate events. As long as the construction industry allows other interest groups to draft construction regulations and legislation, they will fight a wasteful and unnecessary defensive war, one that they are losing thus far.

Employers continue to declare that they are over-regulated and that regulations are impractical, unworkable, and unnecessary. Employers also continue to do the minimum necessary on an individual basis to effect change. For every one construction employer that submits regulatory comment or supports committee participation, there are a thousand that do nothing to effect change.

Proactive or reactive, the choice is up to you.

INDUSTRY SELF-REGULATION

Like it or not, all contractors are not the same. The generic term "contractor" covers the international multi-million dollar construction conglomerate as well as the two-person operation working out of a pickup truck.

The majority of contractors are small employers with fewer than ten employees. Construction at that level is a competitive arena, where the savings of a few dollars means the difference between success or failure. All too often, corners are cut where safety is concerned.

The construction industry needs to ensure that any contractors in the business abide by the rules. The "fly-by-night" operation that does not follow regulations and experiences fatalities hurts the entire industry's image.

The construction industry itself needs to develop the means to clean its own house. Peer pressure is an effective method to accomplish that end.

Chapter 2

Safety and Health Management

INTRODUCTION

Chapter 2 outlines the basic elements involved in establishing an effective safety program. This chapter evolved from a compendium of bits and pieces from construction safety programs in use today. It also relies upon the studies and research that have preceded it and have marked the path we all need to follow.

The figures presented in this chapter can be used as outlines and/or checklists for developing a safety program. Much of the material in the figures can be copied and modified for inclusion into your own company safety program.

SAFETY AND HEALTH MANAGEMENT

Safety and health management starts with a company's commitment to establish an effective working safety program. Planning translates that commitment into reality. To be effective, a safety program must be well-organized. To work, that planning and organization must involve all aspects of the company's business and must encourage the workers to actively participate within that safety effort.

Fortunately, companies do not need to reinvent the wheel each time the decision to start a safety program is established. Companies can take advantage of a multitude of basic safety-program materials available and modify them to fit their particular needs. In addition to the materials contained in this section, safety-program materials are available from the United States Government, construction trade organizations, private entities, and other construction companies.

MANAGEMENT SAFETY POLICY STATEMENT

The company management safety-policy statement outlines the company's philosophy on safety and sets the tone for management's commitment to that safety effort. This policy must be a simple and concise statement of the overall objectives of the company safety program. This policy should assign overall responsibilities for safety in all departments of the company and should be realistic and enforceable.

To communicate the company's commitment to a safer workplace, this policy should be signed by the company's chief executive officer or the highest project management representative.

The sample management policy statements found in Figures 2–1, 2–2, and 2–3 can be used as guidelines in writing your own company policy. They can be used as is, or combined or rearranged to fit your company's specific needs.

Basic Elements of an Effective Safety Program

The Occupational Safety and Health Administration and other organizations have examined companies with effective safety programs and outlined those elements common to an effective safety effort.

1. *Fit your safety program to the size of your business.* Not all construction firms perform the same type of construction work, have the same needs, or have the same number of employees. In fact, the work and the number of employees will differ from job to job and even from job phase to job phase.

   A large company may need a more involved safety program than a smaller firm. A company with a relatively stable work force has a different safety situation than a company with a large employee turnover. The first consideration in developing a safety program is to assess the company's safety needs and risks.

(COMPANY NAME) CORPORATE SAFETY POLICY

The management of (company name) construction company considers no phase of its operations or administration of greater importance than accident prevention. To accomplish this goal it is therefore necessary that an effective and understandable safety policy be stated and enforced.

This company places the responsibility for workplace safety at all levels of management and on each employee. Each member of the company team must work toward achieving the goal of a safe and healthy workplace.

It is in the interest of safety that this company dedicates itself to providing the highest levels of performance in safety, fire protection, and occupational health consistent with OSHA regulations and nationally recognized standards.

Safety shall be an integral part of each job and each employee shall be responsible for the safety phase of his work just as much as any other phase.

The success of (company name) safety program requires the combined efforts of management, supervision and employees. We want our operations to be among the safest in the construction industry. That goal can only be achieved if every person contributes to this team effort.

Sincerely,

Company CEO

Figure 2–1 Sample Safety Policy

(COMPANY NAME) SAFETY POLICY

It is the policy of (company name) to perform all work with the highest regard for the safety of all our employees.

This company will provide a safe and healthy workplace, abide by all regulations as they apply to our industry (as set forth in federal, state, and local standards), and exercise good work practices as dictated by locations and circumstances.

Safety is of the utmost importance in the performance of all operations and must be an integral part of each work task. Safety should never be neglected because of undue haste.

No job is so important or service so urgent that it cannot be performed safely. It is important that all (company name) employees recognize their responsibilities in incorporating safety into their daily work.

Sincerely,

Company CEO

Figure 2–2 Sample Safety Policy

Then, you can modify an existing program or develop materials to address your specific safety needs and situation. Regardless of the extent of your company's safety budget, you can utilize the basic elements outlined in this chapter to establish, modify, or improve your safety effort.

Many of the ideas, like getting your top management

(COMPANY NAME) SAFETY POLICY STATEMENT

The management of (company name) has committed itself to provide a necessary active leadership and complete support in order to develop and maintain:

- A company safety program designed to prevent human suffering, pain, and economic loss from workplace accidents, injuries, or property damage.
- A company program to provide insofar as possible a workplace free from recognized hazards by adherence to federal, state, and local safety regulations and standard industry safe work practices.
- A company work force aware of the workplace hazards that confront them and aware of their safety responsibility to themselves, their fellow workers, and the company.
- A company program to encourage the incorporation of safety into each phase of construction and each work operation and to ensure the security, protection, and well-being of personnel and property at all our work sites.

The company supports safety to the fullest. We call upon each employee to cooperate fully in this safety effort.

Sincerely,

Company CEO

Figure 2–3 Sample Safety Policy

actively involved in the safety effort, are attitude changes that any company can implement for little direct cost.

2. *Management commitment to safety.* A safety effort without the full support of the company's top management will meet with only limited success. Your company should have a written safety policy that is clear and easy to understand. That policy should outline the company's

belief that safety takes precedence over other job-site considerations.

It is foolish to cut corners and take unnecessary risks to get the job done. One accident can turn a small gain into a large loss through increased medical costs, insurance costs, and litigation.

Management must publicize their safety policy to all employees. A well-written policy cannot be effective if no one knows about it.

Management must actively support the safety effort at all levels. Words are fine, but actions are better. When management routinely enforces the company policy and takes action to identify and control hazards, everyone gets the clear message that safety is important. If the management's commitment to safety is lax, then employees will not take safety seriously. Safety and production do not have to be opposing forces. A job can be done safely, on time and within budget.

3. *Clearly defined safety responsibilities.* Your workers can't read your mind. Clearly defined responsibilities and authority for safety and health are needed for each level of management

   Safety should be an integral part of every work task and used as part of the overall evaluation of a supervisor's job performance.

   When supervisors know that safety is one of the yardsticks used to measure their job performance, they become more involved with the company safety effort. Unfortunately, many companies give their line supervisors mixed signals. The company preaches safety, and has an elaborate safety program, but does not consider safety important enough to be used as a job performance criterion. That attitude allows safety to be ignored at the site level because it is perceived by the supervisor as not as important as other factors, like budget or schedule.
4. *Safety programs need money.* A company with a good safety program will ensure that adequate funds are budgeted and spent on the company's safety effort. Often, in construction, money is budgeted for safety, but due to contract wording, it is rarely spent on safety efforts.

There may even be incentives in the contract wording for the project management to return any budgeted but unspent funding back to the construction company or project owners. Then the contract can appear to be completed under the projected budget.

A viable safety effort requires staff and equipment to carry out its functions. Safety expenditures can best be thought of as a preventative maintenance type of funding. Every accident that does *not* occur, every claim that the company is in a strong position to defend against, and every citation *not* issued to your company are safety successes.

A good safety record also means the ability to bid more competitively. With a good record, you pay less for insurance costs. When self-insured, you place less demand on your payment pool.

Owners are now looking at company lost-time accident data and OSHA citation records to decide on a contractor's eligibility to bid on their work.

In addition, a good company safety record means good public image and good public relations.

5. *Leadership by positive example.* The most effective way to lead is by example. The extent of management's commitment to the company's safety effort shows. Your employees take their cue from you and your actions. When safety is only lip service, they know. What kind of example do you set for your employees?

   If management does not set a positive standard of safety behavior for all employees, why should the employees be expected to treat safety seriously?

6. *Open communications.* When the lines of communications between management and employees are not open, management loses contact with its most vital source of expert, job-site specific information. No one knows the problems and conditions of the job site better than the people working there.

   A good management safety program encourages workers to participate actively in identifying and controlling workplace hazards.

7. *Hazard identification and assessment.* Periodic safety

inspections and audits are an important element of a company safety effort. Good safety programs use formal and informal inspections.

Informal inspections can be made on a daily basis by supervisors and workers simply taking the time to examine their own work area. Any problems found are reported and corrected before any injuries result.

Your company should have a policy that encourages workers to report problems to the project management's attention for correction without fear of reprisals. This allows employees to participate actively in the company's safety effort.

Formal inspections can be conducted internally, by the company's safety personnel or by outside services such as your insurance company.

Inspections are important in reminding your line supervisors of your company's interest in maintaining the safety effort. Therefore, they serve to keep everyone on their toes. They also help in documenting your company's safety efforts.

Formal safety inspections and audits should always include a written report. Your company policy should also require a company abatement report for every inspection report.

Failure to document abatement will allow your inspection records to be used against the company in a litigation. It's frustrating to have an opposing attorney cite your own inspection records as proof of your company's "callous disregard for the safety of your workers," because you cannot prove that the conditions reported have been corrected.

Why bother with written reports in the first place? Documented inspection and abatement records show your company's commitment to safety and the existence of a viable company safety program.

8. *Active employee participation.* Enlisting your workers' participation in company safety efforts multiplies the input of information at all levels and increases worker morale.

Employees should be encouraged to report unsafe conditions to their supervisors. Reports, however, are not enough. Actions must be taken whenever such notifications are received. Management's response to potential hazards is important. Response eliminates hazards and provides a clear illustration of the company's commitment to safety for the workers.

Employees also need a forum to voice their opinions and concerns for safety and health on the job. Regularly scheduled safety meetings with employee representatives and management help to "clear the air" and provide a means to continually reinforce the company's safety message.

When employees feel that the company cares about their safety and see enforcement of safety policies, they become much more safety conscious themselves.

9. *Planning for safety.* Planning for safety starts in the bid process and continues until the last workers complete their tasks and leave the job site.

   If safety is left to chance or considered an afterthought, it will not be effective. Safety planning involves developing written guidelines for all employees to follow.

   Safety procedures, rules, and plans cannot be uniform or clear when they are unwritten and rely on word of mouth. To be practical they need to be written and taught to and understood by all employees.

   Planning for disasters and emergencies saves lives, money, and time when crisis situations occur. Figure 2–4 and Figure 2–5 provide a project safety planning checklist and an emergency planning checklist. Additional materials on emergency planning can be found in Chapter 7.

   Figure 2–6 (Project Safety Paperwork Checklist) and Figure 2–7 (Protective Materials and Equipment Checklist) can assist in organizing and planning your project safety needs. These figures can also be incorporated into your company's safety manual.

10. *Employee disciplinary programs.* In addition to a

The following items should be considered when setting up your project safety program:

a. Will there be a full-time safety person assigned to this site?
b. Are a copy of the following on site:
   —Company Accident Prevention Manual
   —Hazard Communication Program
   —Drug and Alcohol Policy
   —Emergency Plans
   —1926 Construction Standards
   —OSHA Jobsite Poster
   —OSHA 200 Log
   —Accident Report Sheets
   —Emergency Phone Number Poster
c. Are on site and community medical/emergency services available and ready?
d. In underground operations, are Mine Rescue Unit(s) available and ready?
e. Is the workplace communications system in place?
f. Are workplace security measures—fencing, lighting, posting operations—completed?
g. Are insurance programs on line and all forms ready for claims?
h. Is there a perimeter survey of utilities, streets, and structures?
i. Are clean up and waste disposals scheduled?
j. Are public safety aspects in place and are there safe accesses for site traffic flow?
k. Are there electrical assured grounding or GFCI programs in place?
l. Is contact established with local emergency service, fire, police, and rescue authorities?
m. Are sanitary facilities and drinking water available as per OSHA Regulations?

Figure 2–4 Sample Project Safety Planning Checklist

written safety policy, a company safety manual, and emergency planning, a written procedure for dealing with employee disciplinary actions is necessary.

A sample written employee disciplinary policy is contained in Figure 2–8. This sample policy can be used as a guideline.

a. Does the site need a rescue or first-aid team?
b. Are community ambulance, fire, rescue, and hospital arrangements made?
c. Are first-aid supplies on site?
d. Are adequate fire protection systems in place, such as signs, fire extinguishers, and storage facilities?
e. Is the site crisis-management team prepared?
f. Have security arrangements been made for the work site including fencing, signs, on-site security, and written orders for security personnel?
g. Are written emergency and evacuation plans maintained on site and distributed to all employees?
h. Has a schedule of emergency training sessions and practice drill been arranged with employees and community emergency response units?
i. Are communications systems installed and available for emergency use?

Figure 2–5 Sample Emergency Plans and Procedures Checklist

Post and make readily available:

a. OSHA posters, Workers Compensation Notice, EEO Notice, minimum wage notice, unemployment poster, safety signs, notices, and a copy of the Company Safety Policy.
b. Required maintenance and testing records for equipment.
c. Written respiratory protection program.
d. Required permits—burning, welding in confined space, etc.—where applicable.
e. Toolbox meeting forms and report forms.
f. Needed warning signs—No Smoking, No Trespassing, Hard Hat Area, etc.
g. Proof of training for power tools, first aid, lasers, chemical use and storage, etc., as needed.
h. Employee safety orientation and safety training logs.
i. Material Safety Data Sheet file.
j. Employee Notice of Unsatisfactory Performance forms.

Figure 2–6 Sample Project Safety Paperwork Checklist

a. Hard hats
b. Safety glasses, goggles, and face shields
c. Approved respirators
d. Boots
e. Rain gear
f. Hearing protection
g. Self-rescuers
h. Knee pads
i. Gloves
j. Fall protection
k. Perimeter guarding materials
l. Safety nets
m. Environmental monitoring equipment
n. Flashlights
o. Emergency barricades, reflective vests and flags for traffic control
p. Storage containers for flammable and combustible materials
q. Tags for defective tools and equipment
r. Approved ladders and scaffold equipment
s. Any specialized safety equipment
t. Hazardous chemical spill materials

Review all work operations periodically to determine if additional safety materials or equipment are necessary. Store and maintain safety materials, equipment, and supplies properly to prevent damage.

Inspect all equipment, tools, supplies, and materials prior to each use, and in accordance with applicable federal, state, and local regulations.

Figure 2–7 Sample Protective Materials and Equipment Checklist

11. *Safety training.* Safety training and orientation are necessary elements of any effective safety program. Supervisors and workers must understand the company's safety policy and procedures and the hazards associated with their work.

    When employees first arrive on-site, a safety orientation training program should be provided. That training session can cover the company and project safety policies, safety regulations, site orientation, personal protective equipment, and OSHA required training.

(company name) has instituted this disciplinary program as one part of the overall company Accident Prevention Program. Part of the supervisors' safety responsibility is to monitor the performance and actions of their crew members and correct unsafe behavior and violation of safety regulations and company safety rules.

Supervisors have total discretion to issue employees a notice of unsatisfactory performance. Remember that this is not a step to be taken lightly. This notice will be used when a worker fails to respond to verbal warnings of violations and unsafe behavior, or when repeat violations or actions occur.

To provide a fair and equitable forum for all employees, any employee adversely affected by a disciplinary action may take the following steps:

1. Discuss the action with the issuing supervisor.
2. Along with an authorized employee representative, discuss this action with the Project Superintendent. Employees will have the opportunity to present any relevant evidence in their behalf. The Project Superintendent may affirm or modify any disciplinary action.

When issuing a notice of unsatisfactory performance to an employee, the issuing supervisor has two options: refer the situation to the Project Superintendent and/or suspend the employee pending further action(s). Any suspension requires a copy of the notice sent to the Project Superintendent.

When issuing a notice, the supervisor should explain the circumstances to the employee and provide the employee with a copy of the notice. Be sure to fill out the notice completely and provide detailed information on the reasons for the notice issued. Attach all supporting material to the notice copy provided to payroll, personnel, or the Project Superintendent.

To assist supervisors in preparing a notice of unsatisfactory performance, consider the following situations in which such a notice is justified:

1. Dishonesty
   a. Theft
   b. Falsification of reports or employment records
   c. Willful damage to company property or equipment

Figure 2–8 Sample Employee Disciplinary Program Notice

2. On-the-job problems
   a. Intoxication
   b. Drug policy violations
   c. Fighting
   d. Horseplay
   e. Failure to report accidents
   f. Abusive or threatening language
   g. Sleeping on the job
   h. Possession of a weapon or firearm
   i. Violation of safety standards
   j. Violation of company policy
3. Attendance problem
   a. Unexcused absence
   b. Chronic lateness
   c. Repeated absenteeism
   d. Leaving work without permission
   e. Other
4. Job performance
   a. Carelessness
   b. Neglect
   c. Poor quality
   d. Insubordination
   e. Failure to correct

Figure 2–8 *(Continued)*

Periodic safety training sessions held with each crew introduces new procedures and reemphasizes safety training. On most construction sites, this is accomplished with a weekly toolbox talk meeting held by the crew supervisor.

Remember, training is more effective when it relates to the conditions and injuries occurring on that job site.

Supervisors need periodic safety training to reinforce safety procedures. Additional training reaffirms the company's commitment to safety and allows a review and discussion of specific job site problems and concerns.

Whenever possible, construction companies should make use of outside sources for additional training programs and sessions. In your community, you will find many potential sources for safety training, such as your insurance company, trade associations, community services and governmental agencies, local chapters of the National Safety Council, professional safety associations, and private safety and health consultants.

12. *Periodic safety performance reviews.* A monthly review of the project safety record including accident statistics, reports of injuries, and results of safety inspections is a

valuable safety tool. Such a review focuses attention on the safety effort and can pinpoint those problem areas that need further safety attention.

It is important to know the causes of accidents so attention can be directed at controlling them. We must consider the cost of accidents when assigning safety resources. A rash of minor accidents that have minimal costs is secondary to a serious accident or a recurring injury with a high cost factor.

Upper Management and Safety

Leadership flows from the top down. Top management's commitment to safety—or lack of commitment—will set the tone for the rest of the company. Top management must take every opportunity to become involved in its company's safety effort.

What can your company management do to create a positive effect on your safety effort? An analysis of companies with exemplary safety records and successful safety efforts shows some common traits shared by those companies' top management.

1. *Top management has direct knowledge of the details of the company's safety program.* Managers of companies with successful safety efforts are aware of all aspects of their company's safety program. They receive and review monthly reports on safety performance.

   Those reports rank individual project safety performance by accident statistics and costs. This allows for a direct comparison with the project's past safety performance and other similar projects. Insurance company loss run reports are valuable tools for assessing this type of data. Construction managers should work closely with their insurance representatives to make full use of the services provided by the insurance carriers.

   Managers of companies with successful safety programs have an open line of communication between project management and the home office. They make it part of their business to show their commitment to the company safety effort. They work to remove layers of bureaucracy between themselves and their employees that impede open communications.

2. *Management emphasizes the safety message.* They take the opportunity to make comments about the company safety effort and the company's commitment to safety.

   Top management can increase safety awareness by taking the time to discuss safety along with costs and schedule concerns whenever they are on site. Awareness of top management's concern for safety reduces the tendency to place production concerns above all else and lets employees know that this company considers safety and production equal partners.

   Successful safety managers also take the time to give positive and negative feedback to employees on safety performance and issues.

   Top management in companies with exemplary safety efforts expect project managers to be knowledgeable about all aspects of their projects' safety program and record. That expectation ensures that project managers will be informed of safety-related issues and problems and more safety conscious.
3. *Safety used as part of performance evaluations.* A company committed to safety uses safety performance as a measurement in employee performance evaluations. Employee promotion and/or other recognitions take into account the employee's record and attitude on safety as part of the overall evaluation process. When a company gives "lip service" to safety and rewards employees for poor safety performance, it is ensuring that employees will not take safety seriously.
4. *Separate safety budgets and accident costs.* Companies with good safety efforts allocate safety expenditures at the corporate level and assess accident costs at the project level.

   Involving the company hierarchy in the safety budget allows the company's top management to become familiar with the dollars and cents involved with safety. Safety, like any aspect of business, can positively or negatively affect the company's bottom line.

   Assessing accident costs at the project level places the responsibility for safety performance with project management, those individuals with direct day-to-day impact on safety performance. It also serves to keep top management informed and to help pinpoint potential safety problems.

5. *Talk about safety.* Companies with positive safety results emphasize the need to constantly communicate the safety message. It's simple—if the employees don't get the message, it can't have any effect.

   Top management needs to communicate positively about safety and to keep reinforcing the message that this company believes that safety is important. There are thousands of ways to keep talking about safety, from company correspondence and publications to informal meetings and the use of commercial safety materials.

6. *Implement safety planning.* Companies with successful safety efforts recognize the value of construction planning. Safety, like any area of construction, benefits from planning work procedures to recognize potential dangers. This ensures that workers are properly prepared and equipped to do the job safely and helps minimize accidents.

How Effective Are Safety Bonuses and Incentives?

The wider an incentive distribution, the more impact incentives have on safety. Incentives that are confined to project managers only do not have a positive effect on the overall safety record of a project. Most effective safety incentive programs concentrate on encouraging a team spirit and highlighting individual recognition.

Safety incentives and bonuses that center on personal monetary gain sometimes provide more of an incentive to "fudge" the accident records than to promote safety. It is also misleading to concentrate safety awards or recognitions strictly on lost-time accident rates.

The overall occurrence of accidents on a work site provides a better picture of that project's safety efforts.

Many projects with zero lost-time accident rates on paper actually end up losing money. Lost-time rates do not tell the complete accident story. A project with a good safety record on paper may end up with excessive costs for a high number of supposedly minor, but costly, accidents and delayed litigation claims.

Incentive and bonus programs can include a wide segment of

the work force, giving each worker a chance to receive some type of ego recognition for a job well done. The use of positive reinforcement, affecting a wide group of employees, is a more effective safety strategy than a bonus or award available to only a few.

Project Management and Safety

Top management may set a company's safety policy, but field management enacts it. The best corporate safety policy is ineffective unless it is properly implemented in the field.

Effective safety managers on the project level share similar traits.

1. *Care about people.* These managers care about the safety of their employees and have earned employee respect. They are able to deal effectively with people and have the ability to resolve conflicts with a win-win type of strategy. Under that approach, they rely on mutually beneficial compromise to solve problems. They exhibit more empathy and concern for the welfare of their employees than managers with poor safety performances.
2. *Display strong safety attitudes.* Field superintendents and project managers with a good basic understanding of the principles of work-site safety and a belief in effective safety programs have fewer accidents and injuries on their jobs. Good safety managers see safety as more relevant to their jobs and believe that the company uses safety as a yardstick for company advancement.
3. *Incorporate job safety planning.* Managers that routinely incorporate safety considerations into job planning recognize and eliminate hazard potentials before they result in injuries. Their planning includes considerations for the necessary safety equipment and proper safe work procedures for each work operation.
4. *Take part in the safety effort.* Managers with good safety records involve themselves in the day-to-day safety routine of their projects. They participate in safety meetings and inspections and maintain daily contact with the field personnel.

5. *Maintain open communications.* Managers with exemplary safety records are viewed by their employees as being concerned with safety. They are often described as accessible when safety problems arise, concerned about conditions on their jobs, and ready to participate in finding solutions to safety problems. They also maintain close home office contact and feel that they can discuss safety concerns at top management levels within the company.
6. *Know about site safety.* Managers who take the time to be informed about safety conditions on their jobs have better safety records than managers who ignore safety considerations. Like the use of "leadership by positive example" cited under Basic Elements of an Effective Safety Plan, concerns for safety exhibited by field management has a positive effect on employees' workplace safety.

   Project management takes its cues from the emphasis placed on safety by upper management. If project management appears unconcerned about site safety, safety concerns on the job suffer.

Supervisors and Safety

Crew supervisors' attitudes and actions on safety affect the safety attitudes of their crews. The traits found in crew supervisors with good safety records include:

1. *Viewing safety as worthwhile and workable.* These supervisors support the company safety effort and understand the relationship between safe and productive work.
2. *Leadership by example.* Safe supervisors follow the company safety rules and present a positive example of safe work procedures and attitudes. They expect workers to follow their example and obey all safety rules and regulations and they correct workers' unsafe behavior whenever they see it.
3. *They integrate safety into all phases of the job.* Safe supervisors tailor the work in progress to the conditions on the work site. They control their work crews and workplace conditions to minimize risks.

4. *They do not ignore near-miss occurrences or dismiss minor injuries as unimportant.* These supervisors take the time to determine why a minor accident or near miss occurred. They view these occurrences as warning signs and an opportunity to prevent a more serious injury from occurring.
5. *They care about their crew and take pride in their work.* Safe supervisors take the time to assess their crews' strengths and weaknesses. They handle new people on their crews differently than more experienced workers. New personnel are paired with a more experienced crew member. They are told what is expected of them and are watched to assess their abilities.
6. *They have control over themselves.* Safe supervisors handle themselves well under stress situations. They analyze problems and work toward solutions rather than placing blame. They act as professionals and deal with people in a positive manner.

SAFETY RESPONSIBILITY

As with all other aspects of business operations, the success of your company's safety program is dependent on management's assignment of responsibility to specific individuals.

Management must make clear assignments of responsibilities to members of the project safety team for those members to understand their roles in the safety effort.

Established responsibilities allow line management and individuals alike to be held accountable for results in the Accident Prevention Program.

Outlined below are the basis responsibilities for the members of your accident prevention program.

Management Responsibilities

Loss control must be an integral part of proper and efficient management. It is a well-recognized fact that safety performance and costs can be controlled. They are as much a part of the overall project costs as production, materials, and quality control.

Management responsibilities include the following:

1. Impressing upon all supervisory personnel the responsibility and accountability of each individual to maintain a safe and healthy workplace.
2. Providing employees with the necessary safety training in all facets of their work operations.
3. Distributing information, reports, accident data, and changes in health and safety regulations or codes that pertain to company operations.
4. Providing all supervisors with copies of written appropriate rules and regulations.
5. Incorporating safety when planning for all phases of the project to maximize the use of engineering and administrative controls in the overall accident control program.
6. Continuing to monitor all aspects of the program for overall effectiveness, necessary assistance to field personnel, and compliance with company policy and safety regulations.

PROJECT SUPERINTENDENT RESPONSIBILITIES

The Project Superintendent is responsible for the active administration and control of all aspects of the workplace safety program. The superintendent can most effectively reduce the occurrence of accidents and improve safety when the company's corporate safety program has full support. Companies demonstrate that support in many ways, i.e., by including safety as an integral part of job planning and by supporting the development and use of safe work practices by all levels of project employees. The duties and responsibilities of the Project Superintendent should include but not be limited to:

1. Planning and requiring that all work operations be done in compliance with established safety regulations and safe work practices.
2. Appointing a competent person (as defined by OSHA) to assist in the project safety programs.
3. Ensuring the availability of all necessary personal protective equipment, job safety materials, and first-aid facilities.
4. Ensuring that all new employees are properly instructed

in safe work practices, hazard recognition, and OSHA-required training, and conducting safety-training programs prior to field assignments.

5. Ensuring the maintenance of all safety record keeping and documentation. This includes accident reports, toolbox meeting attendance, safety inspections, audit reports, employee training records, and all OSHA record keeping.
6. Instructing all supervisors about their safety responsibilities prior to initial project activities.
7. Reviewing all lost-time, injury, accident, and property-damage reports with the appropriate supervisors.
8. Ensuring to the greatest extent possible subcontractor compliance with your company's accident prevention policy, and all applicable federal, state, and local standards and regulations.
9. Communicating safety-related information and reports to all employees.
10. Keeping company home offices informed of all safety activities, communications, and problems.
11. Properly treating all injuries and investigating all accidents, injuries, or property-damage occurrences.
12. Implementing all emergency procedures in response to a crisis situation and following all emergency planning procedures and policies until relieved by corporate representatives.
13. Assigning or delegating the responsibilities outlined above to qualified and competent individuals. When delegating responsibilities, the Project Superintendent must monitor these activities to ensure full compliance.

Supervisors' Responsibilities

We must realize that the attitude developed by workers toward safety is a direct reflection of the supervisor. Therefore, all supervisors must pay prompt attention to, and take appropriate action on, employee safety suggestions and report unsafe conditions and practices. The following supervisory responsibilities address these objectives:

1. Ensuring that all employees under your supervision understand their safety responsibilities.
2. Ensuring proper use of all necessary personal protective equipment.
3. Acting without delay to address all hazards and unsafe conditions or actions within the scope of your area of responsibility.
4. Cooperating fully in the investigation of any accidents, taking all necessary measures to prevent a reccurrence, and reviewing causes and possible prevention of accidents with the employee(s) involved.
5. Informing project management of any safety problems that lie beyond your authority to correct.
6. Holding weekly toolbox safety training meetings with the workers under your supervision and maintaining proper toolbox documentation and submitting them as directed.
7. Caring for all injuries immediately and reporting all accidents properly, a report should follow all injuries involving lost time, a doctor's care, or damage to company property, equipment, or facilities above a potential cost set by the company (usually $250.00).

Employee Responsibilities

The employee plays a vital role in your company's safety program. Without the full cooperation of each employee, your company cannot achieve the full potential of your accident-prevention program. Employee responsibilities include:

1. Consistently observing work conditions, operations, equipment, and tools for the purpose of accident prevention.
2. Complying with all federal, state, and local regulations as well as safe work practices and company safety rules.
3. Using all necessary personal protective equipment, safety equipment, and safe work practices required for the safe performance of the job.
4. Correcting any unsafe work practices within the scope of the work operations and reporting any unsafe work actions or conditions to the supervisors.

5. Examining tools and equipment before use and advising supervision of any defects or problems; do not use a defective tool or piece of equipment for any reason; do not use tools or equipment improperly or in a manner for which the tool or equipment was not designed.
6. Stopping work if conditions present an immediate danger to life, limb, or property and reporting such conditions immediately.
7. Availing yourself of all company and industry sponsored safety training, information, or programs.
8. Assisting other employees in performing their work in a safe manner by informing them of potential problems, unsafe conditions, or unsafe actions.

Subcontractors' Responsibilities

Require all subcontractors to adhere to your company's accident prevention program. In addition, subcontractors responsibilities should include:

1. Abiding by all safety rules of the owner and other project contractors.
2. Notifying all other contractors when actions or activities of your work operations may present a safety or health problem to other contractors' employees.
3. Informing your company whenever you will be on-site.
4. Informing your company of all accident or injury claims.
5. Reporting and observing unsafe conditions or actions to the appropriate authority.

Chapter 3

The Worker Component of Accident Causation

THE WORKER COMPONENT OF ACCIDENT CAUSATION

Work-site accidents occur through three primary mechanisms: unsafe conditions, unsafe acts, or acts of God. Acts of God are outside the realm of man's control and unsafe conditions are addressed by volumes of OSHA regulations. Why then do we ignore the unsafe acts component in the application of safety regulations in the United States?

When the Occupational Safety and Health Act was passed in 1970, Congress appeared intent on addressing both the employer and the worker responsibilities for safety compliance. Section 5a of the OSHA Act requires employers to provide a workplace free from recognizable hazards. Section 5b requires workers to obey all safety regulations. This seemed to be an equitable partnership, with the worker and employer sharing workplace safety responsibilities.

In 1976, the courts interpreted the OSHA Act in a far different light. They ruled that the language used by Congress in the Act places an obligation for safety only upon the employer. OSHA, therefore, in an internal policy decision, decided that they had no authority to enforce worker safety compliance. OSHA's actions left only one possible answer in the OSHA equation of job-site safety—no matter what causes an accident, no matter who fails to follow regulations, only the employer is cited for safety violations. In essence, under the auspices of the OSHA Act, workers have no responsibility for their own safety.

Deliberate acts of employee misconduct cannot be the basis of a defense against an OSHA citation, unless the employer can prove that the employer did not condone or permit such employee actions, which is a kind of "the employer is guilty until proven innocent" assumption by OSHA.

The effects of OSHA's policy of ignoring the employee component of accident causation are far-reaching. That policy undermines effective workplace safety efforts, stifles solutions to accident control, and is a factor in the increase of injury liability lawsuits.

Many other nations, like Canada, allow the citation of both the employer and the worker when a workplace safety violation occurs. Although actions against individual workers in Canada are infrequent, that potential is still a viable safety motivator.

Employers in the United States have no such "big stick." Their actions in instances of deliberate safety violations by a worker are limited. There is always the threat of termination of employment for safety violations. However, the realities of a shrinking labor market and the tenets of collective bargaining agreements limit effective use of that option.

Employers in construction cannot always hire enough skilled labor to meet their current job needs. How many times have you passed a construction site and seen a sign advertising jobs? In addition, union agreements sometimes make it difficult to terminate an employee. Termination, in any case, is a limited solution—a solution that only transfers a problem employee to someone else's payroll.

A government truly concerned about workplace safety in the United States should hold employees accountable for their intentional violations of safety regulations. Under current OSHA law, that is impossible.

What we need is a revision of the OSHA Act to make both the employer and worker responsible for safety on the job—a partnership of responsibility, if you will. However, that type of needed reform of the OSHA Act is unlikely. Elected officials do not wish to alienate a large voting block controlled by organized labor.

What role then does the worker component of accident causation play in the number of workplace accidents? The precise answer is not clear.

With the billions of taxpayer dollars spent on American safety and health, we have never conducted a national study to pinpoint the causes of workplace accidents. Yes, OSHA and others do collect data on accident occurrences, but the emphasis has not been on finding out why accidents occur. Given the current state of safety responsibility under OSHA (i.e., the employer automatically being responsible), such a study is unlikely. With the worker component of accident causation removed from OSHA consideration, there has been no impetus to conduct such studies.

However, some sources outside the government clearly indicated that the number of worker-caused accidents are high.

NATIONAL INSTITUTE OF OCCUPATIONAL SAFETY AND HEALTH (NIOSH) DATA

OSHA's own preamble to the electrical safe work practices standards cites a NIOSH study[1] on worker-related accidents. The data presented indicate that 50 percent of workplace electrocutions are due to unsafe acts on the part of workers.

These electrical fatalities, according to OSHA, were not due to faulty equipment or violations of OSHA standards by employers. They were due to an unsafe work practice by experienced electrical workers, workers who should have known better.

Another NIOSH work injury report on chemical burns in 1985[2] found that 58 percent of the burns occurred among experienced employees, workers experienced in working with the materials involved. Over two thirds (69 percent) of those employees injured were not wearing proper personal protective equipment when injured.

A survey of workers by NIOSH about the cause of their own accidents indicated a number of accident causes. The conditions or factors that workers felt contributed to their accidents included: working too fast (22 percent); careless work (12 percent); using equipment improperly (12 percent); upset, under stress, or tired (7 percent); not paying attention (6 percent); and co-worker activity (4 percent).

[1] OSHA proposed standard on electrical safe work practices Federal Register number 2047, 21694.

[2] NIOSH Work Injury Report (WIR) Chemical Burn Survey May 1985.

DUPONT "STOP" PROGRAM DATA

Dupont Safety Services markets a safety management program that is an outgrowth of Dupont's safety experience and research. Dupont's Safety Training and Observations Program[3] (STOP) states that 90 percent of accidents are due to unsafe acts by workers.

This program, geared to identifying and controlling worker unsafe acts, has achieved remarkable results in reducing workplace accidents. With Dupont reporting lost work day incident rates as low as 0.019 per 200,000 exposure manhours, it would seem that given these results, there is merit to this approach of controlling unsafe conditions and unsafe acts!

Our national safety effort turns a blind eye to reality of employee accident causation while accidents continue to occur. How much more could be achieved if a "STOP" type of safety emphasis program was made a national safety imperative?

BONNERVILLE DAM PROJECT

Several individual project studies shed light on worker accident causation. One study, on the Bonnerville Dam Project,[4] reported that seven times the number of work accidents that occurred on this project were due to unsafe employee actions rather than unsafe site conditions.

That study on the Bonnerville Dam project concentrated on how accidents happen. Knowing the cause of accidents allowed the contractor to target areas for special concern and work toward reducing the number of accidents. The Bonnerville study also found that workers' negative attitudes toward safety were a major factor in accident occurrence.

THE STANFORD STUDY

A study by Stanford University[5] indicated that risk taking is often a normal part of human psychology. We sometimes drive

[3] Dupont Safety Training Observation Program S.T.O.P., Dupont Company Safety Services Promo #E-89581.

[4] Dambuilders Plots Safer Worksites, *Construction Equipment Magazine,* July 1979.

[5] A Survey of the Safety Environments of the Construction Industry, Stanford University Construction Engineering and Management Program, Department of Engineering, Stanford University 1980.

too fast or take chances we should not. Risk taking on the job site, however, can be fatal.

The Stanford study found that many workers believe that taking unnecessary risks is an accepted part of the job process. This risk-acceptance attitude leads to carelessness and accidents. The results of the Stanford study show that workers who are likely to have lost-time accidents share similar characteristics. These workers have a negative attitude toward doing their jobs safely, and they accept unnecessary risk and, therefore, do not work safely. Taking unnecessary risks and a poor safety attitude simply makes workers more prone to accident occurrences.

The conclusion of the Stanford study clearly supports the contention that employee actions and attitudes can affect the number and type of workplace accidents.

Employers can and do address this attitude of risk taking through safety education, safety rules, and training programs. However, no employer can supervise each employee every minute of the work day.

Employers simply do not have the control or the resources to ensure that all their workers abide by the rules or work in a safe manner all the time.

Yet, this is the very position they are forced to adopt under OSHA. Employers are the only party that OSHA holds responsible for workplace safety.

BALANCING THE EQUATION

Regardless of which numbers we use, a percentage of accidents are caused by employees. Whether or not employee accident causation is less, equal to, or exceeds the number of accidents caused by unsafe conditions, it cannot be simply ignored. Unfortunately, our country's workplace safety effort under OSHA does just that—ignore obvious possibilities. How can we expect to reduce the occurrence of workplace accidents by ignoring some of the reasons why they occur?

We must balance the equation for safety responsibility to address the causes of all accidents. The reduction of workplace accidents and injuries starts with the fundamental understanding of why and how they occur. Once we have that knowledge, we can treat the causes of those accidents. Without the application of that basic premise, we can never find meaningful solutions to accident prevention.

Safety professionals understand that safety is and always will be a people business. Success in safety programs is achieved through the motivation and education of people. However, a majority of our safety effort and budget is consumed in compliance with regulations and the protection of the employer from citation, efforts that would be better spent in addressing employee risk taking.

Employers today find themselves over-regulated, forced to spend time and money in compliance with safety standards that are not beneficial or appropriate to safety. The regulations seem geared toward creating paperwork and not toward making the work place safer.

OSHA's adversarial relationship with employers and emphasis on enforcement requires employers to conduct business defensively. Employers must protect themselves against the possibility of OSHA fines and OSHA's use of the employer's citation history to issue repeat and willful citations.

Additional pushes for state and local prosecution, and for workplace accidents and injuries, fuel the need for employer self-protection. Meanwhile, we move farther and farther away from finding lasting solutions to our safety problems.

Ignoring the employee component of accident causation has also fed the atmosphere of rising litigation. The employer as the responsible party for workplace safety is always the one cited by OSHA. Once cited by OSHA as "responsible" for a safety violation, that citation is used against the employer as proof of liability.

Citations create a record that follows an employer. Citations are often used as evidence of the employer's apparent lack of safety effort. Citations also suggest the employer had prior knowledge of a dangerous condition or operation. This evidence provided by OSHA citations is often used in employee claims for damages in court proceedings against employers.

Juries convinced that OSHA citations of an employer must indicate deliberate wrongdoing on his or her part return awards regardless of the facts and realities of the case, judgments often based on sympathy and the "David versus Goliath syndrome." The assumption is that insurance companies and employers have lots of money so they will not miss the money, whereas an individual worker certainly would.

A reform of OSHA could restore the balance of safety responsibility between employers and employees. We need the type of reform where individuals are responsible for their own actions.

Workers facing the specter of an OSHA fine for their own violations of safety regulations will be more aware of their own safety responsibilities. Indeed, both the worker and the employer would have a personal stake in ensuring workplace safety compliance.

Not holding workers responsible for violation of safety regulations is akin to citing only the car manufacturer for traffic offenses. How many drivers would obey the traffic laws under those circumstances?

Current accident statistics do not record the underlying cause of workplace accidents and injuries. Federal statistics "bean count" the number of accidents and injuries that occur. The causes of those accidents are left undetermined.

The reason for this oversight is not difficult to understand. When the only possible answer to a workplace accident is to cite the employer, there is no need to look for further answers. This creates a climate for minimal federal effort to understand the underlying cause of accidents.

The government does maintain statistics on the mechanism of injuries. Those statistics deal with the objects involved rather than the root cause of that accident.

A national study to examine the base causes of workplace accidents and injuries would be an invaluable resource for safety improvement. Given, however, the current conditions of the federal safety effort, such a study is unlikely, and we will all be worse off for it.

We are, perhaps, prisoners of the very system we have created—a system sometimes devoid of common sense. Will we continue to spend billions of dollars on safety regulation and compliance while ignoring the reasons accidents occur—a shotgun approach to safety with regulations enacted in the hope of somehow hitting the target, even when the target is unknown?

Chapter 4

Dealing with OSHA

INTRODUCTION

Chapter 4 deals with the Occupational Safety and Health Administration, OSHA, from an employer's perspective. The purpose of this chapter is to provide management with the information they need to develop a company policy for their self-protection while dealing effectively with OSHA inspections.

Nothing within this chapter or the book should be construed as advocating or endorsing noncompliance with OSHA regulations. OSHA is a fact of life that employers may not always like, but must deal with. Compliance with OSHA's regulations is the minimum that an employer must do to provide a safe and healthful workplace for all of his or her employees.

DEALING WITH OSHA

Allowing OSHA free access to inspect your job site is like the farmer inviting the fox into the hen house. It is not good for the farmer's business and very bad for the chickens!

This is because OSHA has chosen to assume the role of a regulatory police officer to American businesses rather than a safety resource and educator. The agency chooses to place the emphasis and the majority of their efforts on enforcement of OSHA standards through workplace inspections.

That enforcement emphasis, coupled with the recent trend by OSHA to issue high dollar fines and the unrealistic requirements

Table 4–1 Employer Conduct in OSHA Inspections

If you decide to allow an OSHA inspection of your work site, take the following steps to protect yourself:

1. Be professional and polite.
2. Review and record the credentials of all compliance inspection personnel.
3. A qualified, properly equipped employer representative should accompany each compliance officer during the inspection.
4. Take detailed notes about the inspection.
5. Take additional samples and photographs of everything so documented by OSHA. These will be in your defense.
6. Do not make statements to OSHA that work against you.
7. Do not agree with statements that place blame or responsibility for violations on the employer.
8. You are not required to answer any questions, demonstrate machines or equipment, or volunteer any information.
9. Do not interfere with the inspection process. Remember your absolute right to limit the scope of an inspection.
10. Abate unsafe conditions in a safe manner.

of many OSHA regulations, forces more and more employers to channel their efforts into protecting themselves from OSHA. Table 4–1 provides the employer with guidelines for protecting itself in an OSHA inspection.

There is a growing sentiment within the construction industry that OSHA is something to be tolerated as long as it doesn't seriously interfere with the company's safety efforts.

Whether you agree with OSHA's emphasis on enforcement or not, understand that the sole purpose of the OSHA compliance officer in conducting any workplace inspection is to gather enough evidence to issue citations against the employer.

Although OSHA is reluctant to admit it, a compliance officer's citation history is a factor for promotions within OSHA. Compliance officers base their issuing of a citation on their observations during an inspection and their knowledge of the industry being inspected. That knowledge of your work operations and your industry may be limited to only a few weeks of OSHA classroom training.

Employers often feel that they run a good, safe operation and, therefore, have nothing to fear from an OSHA inspection. They cooperate with the inspector, answer all questions, and even volunteer information, only to find their cooperation used against them when citations are issued.

Employers have the constitutional right to refuse OSHA access to their workplaces without a legally obtained warrant. Employers also have the right to limit the scope of an OSHA inspection and use the OSHA inspection process as an opportunity to prepare a defense against alleged citations.

Employers need to fully understand their rights under OSHA and prepare their actions far in advance for an OSHA inspection. Table 4–2 provides a review of the employer's rights when dealing with OSHA inspections.

Table 4–2 Employer Rights in OSHA Inspections

1. You have the right to control access to your work site.
2. You have the right to refuse an inspection without a warrant.
3. You have a right to know the scope of the inspection.
4. You have a right to a reasonable inspection process, conducted at reasonable times and in a reasonable manner.
5. You have the right to defend yourself and your company against alleged citations.
6. You have the right to have a company management representative present during inspections.
7. You have a right to avoid undue or unnecessary disruption of your work schedule during an inspection.
8. You have the right to accompany the inspector during his physical inspection of your site.
9. You have the right to disagree with the compliance officer's assessment of an imminently dangerous situation.
10. You have the right not to provide OSHA with the place or work time to interview employees.

OCCUPATIONAL SAFETY AND HEALTH ACT—HISTORY

The Occupational Safety and Health Act of 1970 was enacted by Congress to ensure, insofar as possible, safe and healthful working conditions for all employees.

The Act directed the U.S. Department of Labor to establish and enforce standards regulating safety and health in the workplace.

This created a mad scramble for OSHA to find and adopt an initial group of applicable industry standards within a specified period. In construction, a mismatched set of existing standards, including the Contract Work Hours and Safety Standards Act, The Walsh Healey Act, ANSI Standards, and others, were compiled into the initial Construction Industry Safety Standards.

Since that initial adoption, OSHA standards have been undergoing revision based more on the political climate and external pressures than industry need. Standards are often written by OSHA staff personnel who have a minimum understanding of the industries they seek to regulate.

OSHA does not work directly with the affected industries to develop drafts of proposed standards. The comment period following a proposed OSHA standard often seems to be a place for OSHA to sift through submitted materials to extract those comments that fit OSHA's preconceived regulatory opinions and ignore the rest.

Regardless of anyone's personal feelings toward OSHA, the language of the Act does require employers to follow all the standards set up under OSHA. OSHA is, therefore, a reality of construction life we cannot ignore.

The Occupational Safety and Health Act applies to all employers engaged in interstate commerce. The courts have ruled that if you pay taxes, use a phone capable of calling across interstate lines, or use goods or services manufactured in another state, you are engaged in interstate commerce.

OSHA's jurisdiction covers all of the fifty states, the District of Columbia, Puerto Rico, and all territories under the United States Government jurisdiction.

The only employers exempted from coverage under OSHA are self-employed persons, farms with fewer than 10 employees, state and municipal employees, and any workplace governed under any other federal jurisdiction. The exclusion of government and municipal workers and other groups seems to be at odds with the stated purpose of the Act "to provide a safe and healthful workplace for all employees."

Construction regulations promulgated under OSHA are found in the Code of Federal Regulations, titled 29 CFR Part 1926. These standards cover all kinds of construction, from light to heavy and from residential to commercial. Construction standards have been applied by the courts, even to work sites involving alterations performed on a homeowner's garage.

THE GENERAL DUTY CLAUSE

Section 5 (a) (1). Under the General Duty Clause 5 (a) (l) of the OSHA Act, employers are required to furnish, to each employee, employment and a place of employment free from recognized hazards that are causing or are likely to cause death or serious physical harm to employees.

OSHA defines recognized hazards as a hazard that could have been known or should have been known to the employer because of applicable regulations or standard industry safety practices.

Section 5 (a) (2) of the OSHA Act requires each employer to comply with Occupational Safety and Health Standards promulgated under this Act. The courts consider compliance with OSHA Standards as the minimum acceptable level of safety effort by an employer.

Section 5 (b) of the OSHA Act states: "Each employee shall comply with occupational safety and health standards and all rules, regulations, and orders issued pursuant to this Act, which are applicable to his own actions and conduct."

At first, the requirements of Sections 5(a) and 5(b) of the Act seem to create an equal responsibility for safety between the employer and the workers.

However, later interpretations by various courts would rule that the language of the OSHA Act places a duty for safety only upon employers. Workers are to obey all safety regulations, but OSHA has no jurisdiction or authority over workers and therefore cannot penalize them.

Under OSHA that lack of jurisdiction over workers is deemed not important because employers have the means to control workers' safety violations. After all, they have the responsibility to take disciplinary action against workers who violate safety regulations.

Employers and their management need to clearly understand the fundamental fact that employers alone are accountable to OSHA. OSHA, therefore, directs all of its enforcement actions only against the employer.

The courts have ruled the requirements for workers to obey safety regulations are unenforceable by OSHA and, therefore, are only words on paper lacking any meaning.

Before employers can use worker misconduct as a defense in an OSHA citation, the employers must prove that they took all necessary steps to prevent intentional employee violations of safety regulations.

Employer control over worker behavior is tenuous at best. Collective bargaining agreements in union sectors often impair the employer's ability to take disciplinary actions against an employee.

In open shop areas, disciplining an employee may not solve the workers' underlying safety attitude or motivational problem. In union or open shop work, by firing a worker, one employer may just be passing the problem on to the next employer.

Not holding the individual worker accountable to OSHA for personal violations of regulations and safe work practices is a glaring oversight in OSHA's workplace safety effort. It is also a major obstacle to effective workplace accident control. This "free pass" from safety responsibility for workers makes as much sense as holding the owner of a car, not the driver, responsible for traffic violations.

Understanding the existence of this OSHA blind spot to a worker's safety responsibility helps the employer obtain a better perspective of the entire OSHA process.

Perhaps the first mistake employers make when dealing with OSHA, on the national or regional level, is the assumption that OSHA will be impartial and that their rules will make sense.

Starting with the understanding that OSHA can only deal with the employer component of the complex reality of construction safety eliminates that expectation of logical, unbiased action by OSHA.

There is only one conclusion that OSHA can ever recognize—the employer is at fault!

OSHA has also become an ever-growing and complex bureaucracy, a part of the vast government process. As such, it is subject to random and sudden change based on the political climate and bureaucratic necessities.

OSHA goes through periods of activity and inactivity relative to job-site enforcement and the standard setting process. Currently we are in an active period.

Recent OSHA trends, including the issuing of high dollar OSHA fines, changes in the inspection process, and a new wave of standard setting, make it more important than ever for employers to protect themselves in OSHA inspections.

SELF-PROTECTION FROM OSHA

Some attorneys are advising their employer clients of the need to consider having attorney representation during the OSHA inspection process and of routinely denying OSHA access without a warrant.

This approach may be considered drastic, but it does highlight a serious and growing resentment of OSHA by business. When an OSHA inspection and fines can force a company into bankruptcy and subject management to criminal prosecution, a drastic approach merits serious consideration. The approach that your business takes in dealing with OSHA must be based on expert legal advice and your company's philosophy, situation, and needs. Whatever approach you choose to use must be applied uniformly to all parts of your company.

Employers are often reluctant to require warrants for OSHA inspections because it smacks of implied guilt, the "I've got nothing to hide" concept. Employers need to remember that running a safe job site and compliance with OSHA regulations are often distinctly opposite ideas.

OSHA's waste of employers' time and safety efforts in enforcing ill-conceived and illogical regulations, like the regulation detailing the height for hanging fire extinguishers early on in OSHA's history, are legion.

Citations by an OSHA inspector for alleged violation of OSHA standards are not a violation of law. It is only a statement of that particular OSHA inspector's opinion. It is, in fact, up to OSHA to prove that the violation occurred and represented a danger to the employees' safety and health.

Employers often fail to understand that they do not have to allow an OSHA inspection without a warrant. Exercising and protecting their constitutional rights as a company is not illegal, unwarranted, or an admission of guilt.

The Fourth Amendment to the Constitution of the United States protects private property from unreasonable searches. The courts have ruled that an unreasonable search is any inspection conducted without a warrant, unless the employer or an authorized agent consents to that particular search.

Some companies have chosen to exercise their right to protect themselves from unreasonable searches by taking a hard line and/or by outlining their position in writing. They do this by providing a written notice to each OSHA inspector arriving on site outlining the company's OSHA inspection policy. Figure 4–1 provides a sample of a notice to OSHA regarding inspections.

After deciding whether or not to exercise the company's right to be free from unreasonable OSHA inspections, the next logical

It is (Company Name's) policy to cooperate with any agency of the United States Government seeking to lawfully enforce all laws and regulations under federal, state, and local jurisdictions. That cooperation is pursuant to the safeguards guaranteed under the Constitution of the United States.

Please Take Notice: (Company Name) chooses to exercise our constitutional rights and will not permit, nor have we authorized any employee of this company the right to permit, an inspection or search of this workplace by any OSHA representative unless it is pursuant to a legal and valid warrant.

Figure 4–1 Notice to OSHA Regarding Inspections

(Company Name) does not believe that sufficient probable cause exists for OSHA to inspect this workplace and requests that the company be notified of any application by OSHA to obtain such a warrant so that we have the opportunity to oppose the same. Should the company decide to honor any warrant presented by OSHA, any allowance for such an inspection will be UNDER PROTEST.

The company will allow a warrant inspection only under the belief that it may be held in contempt of court for declining to allow the inspection to begin as stated in the warrant. Nothing within this notice, or in our company's decision to honor a valid and legally executed warrant, will be construed as an agreement to waive our rights to challenge the validity of this inspection or the authorization for its conduct.

This notice to OSHA was given to (Compliance Officer's name), as a representative of OSHA, on (date) in the year 19____ at (time).

(Signature of Company Representative)

Figure 4–1 ***(Continued)***

step for employers is to develop a written policy for dealing with OSHA inspections. That policy gives all supervisors guidelines for their conduct during inspections. It also teaches management how not to actively assist OSHA inspectors in collecting evidence of employer noncompliance with OSHA regulations.

The sad fact is, unfortunately, that in their efforts to cooperate with OSHA, most employers assist in convicting themselves of OSHA violations.

Don't think that OSHA doesn't appreciate your help; they count on it.

THE EMPLOYER AND OSHA INSPECTIONS

Employers need to understand their obligations and their rights under the Occupational Safety and Health Act. Company management on site needs to know what to do and how to act when OSHA knocks at the door. When the OSHA inspector arrives at the site, the time to learn and prepare has expired.

Employers need a written OSHA policy because the job superintendent or the company official on site may not be trained to deal with the compliance officer, the inspection process, or protecting the rights of the employer and the company.

Without prior knowledge and preparation, an employer may waive his or her legal rights and undermine their protection. The unprepared employers find themselves in a poor position to contest alleged OSHA citations.

OSHA's Right to Inspect

OSHA has the right to inspect those work sites under its jurisdiction without advanced notice. In fact, anyone providing advanced notice of an inspection is subject to OSHA fines and criminal actions.

Employers, under the "Barlow Decision"[1] have the right to refuse OSHA access to the site and require that OSHA obtain a warrant before conducting any work-site inspection.

To obtain that warrant, OSHA needs to show to the court that the inspection is justified under the OSHA Act. A justified inspection is one routinely scheduled, based on an employee complaint, or part of the follow-up inspection process. The procedures that OSHA must follow to obtain a warrant are time-consuming and burdensome. The employer as the inspected party has the right

[1] Marshall v. Barlow, Inc. (436 U.S. 307, 1978).

to oppose OSHA's application to the court for an inspection warrant.

OSHA's Field Operations Manual, "the rules for OSHA compliance officers," instructs them to immediately cease inspection operations and contact their area office when an employer refuses them site access.

The Compliance Officer will notify the OSHA Office Area Director. The Director, in turn, will notify the Regional Administrator. The Regional Administrator will consult with the Solicitor's Office regarding further actions. Any one of these persons may decide that a warrant is no longer advisable or cost effective.

If an employer or an employer representative elects to refuse OSHA permission to inspect without a warrant, he or she needs to take another step for self-protection. You should place OSHA on written notice of your position and ask them to provide you with notice of any application to the court for a warrant to inspect your work sites. There is no required form for this notification. A written letter to OSHA and the court will accomplish that notification. A sample of a notification letter to OSHA and the court is included in Figure 4–2.

Date:

TO: CLERK OF THE COURT

(Address of court with jurisdiction for OSHA cases)

Subject: OSHA APPLICATION FOR A WARRANT TO INSPECT

(COMPANY NAME)

On (date OSHA attempted inspection), the Occupational Safety and Health Administration attempted to conduct a warrantless inspection of (Company Name's) work site located at (address of project).

Figure 4–2 Notice to the Court and OSHA

We declined to permit this inspection by (name and address of OSHA office attempting to conduct an inspection) because we feel that probable cause for such an inspection does not exist under the rule of Marshall v. Barlow, Inc. (436 U.S. 307, 1978) or under any other basis.

We have informed OSHA in writing of our objections to this warrantless inspection on (date and time of notice).

We request notification in advance in the event that OSHA attempts to file an application to obtain any warrant to inspect our company facilities or work site so that we may present our reasons why probable cause for such issuance does not exist.

(Signature and Address of
Company Representative)

cc: OSHA
file
Congressional and State Representatives

(Be sure to send a copy to the OSHA Area Director's office, where the citation was issued, to place OSHA on notice.)

Figure 4–2 ***(Continued)***

OSHA INSPECTION UNDER A WARRANT

Whenever OSHA arrives with a warrant in hand, the first thing for an employer to do is to ask the OSHA inspector(s) to wait. Then, make a copy of the warrant and carefully review it before allowing OSHA to continue. The warrant will specify the extent and time for the inspection period. It will also detail the scope of OSHA's inspection and the address of the court that issued the warrant. It is not a crime to delay OSHA. You may also use this time to consult with your company's main office or legal council.

Employers have the right to refuse to honor a warrant presented by an OSHA inspector. In essence, you may object to the validity of the warrant or refuse to allow OSHA to conduct an inspection even under a warrant. Employers have the right to file a motion against OSHA to squash or void the warrant with the court of issue. OSHA will then likely go back to court again and request a court order to hold your company in contempt for refusing to honor the warrant.

A scheduled hearing will be held where you can defend your actions. If you win, the warrant will be voided. If you lose, the inspection will continue, but with a legally defined direction and scope.

Why are some companies going through the trouble of requiring a warrant? A warrant limits OSHA's rather broad inspection power. Without a warrant, OSHA has much more latitude to decide what it wants to look at in your workplace and what workplace violations it chooses to cite. Under a warrant the purpose, place, and scope of the OSHA inspection are specified. The choice to honor the warrant or not is up to each employer.

Whether you choose to exercise your right to honor a warrant or not, employers should always take the opportunity to object to the warrant's validity. Employers can inform OSHA in writing that they do not consent freely and will allow the inspection only under protest. That protest protects your right to contest the inspection and the warrant's validity at a later date.

You have the legal right under the Barlow Decision to delay OSHA's inspections until you have received a copy of the warrant and the warrant's supporting documents and references. You further have the right to thoroughly read the warrant and have a clear understanding of the scope and purpose of OSHA's inspection. In exercising that right, you may question the inspector and require clarification of any warrant language you do not understand.

Employers have the right to know the purpose for and scope of any OSHA inspection. When the inspection is due to an employee complaint, employers have the right to know the nature of that complaint. They also may request a copy of the alleged complaint. In most cases, OSHA will seek to protect the identity of the person making such a complaint.

OSHA INSPECTIONS

As we stated earlier, OSHA conducts inspections of your workplace for the purpose of gathering evidence of alleged violations of their regulations. This evidence will form the basis for citations issued against the employer and possible convictions of the company as an OSHA violator.

There are two times when OSHA inspects your workplace—those occasions when OSHA obtains a warrant, and those times where your actions seem to say, "OK, OSHA, come and get me." The value of a warrant is that it defines OSHA's authority to inspect your workplace. The warrant clearly outlines the extent of OSHA's inspection powers. For example, if the warrant does not specify OSHA's right to review your records, you have the right to refuse OSHA access to those records.

Even if you decide to allow OSHA to inspect without a warrant, you have the right to limit the areas where the inspector can conduct an inspection or terminate the inspection at any time.

You have no obligation to assist the OSHA inspector's efforts to gather evidence against you. In fact, you would be foolish to do so!

A prime rule for the employer in all OSHA inspections is to request OSHA to define its inspection purposes and limitations, as well as putting into written form any verbal agreements regarding the scope of an inspection. If the OSHA inspector declines to provide a written agreement regarding the scope of an inspection, you have no obligation to permit an inspection without a warrant.

OSHA's New Inspection Process

OSHA schedules inspections through local office scheduling programs and lists, using local high-risk industry targeting and special interest criteria. For example, trenching, excavations, and lift-slab operations are currently targeted for special inspection efforts.

OSHA currently schedules construction industry inspections through the use of the "Dodge Report of Construction Potentials." McGraw-Hill, publisher of the Dodge Report, provides construction project information for OSHA to develop their area office inspection lists. If you are not listed with the Dodge Report, your chances for an OSHA inspection under the current targeting program are decreased.

In 1988 OSHA changed a long standing policy on workplace inspections. OSHA used to provide an exemption from routine scheduled inspection for employers with better-than-average safety records.

All employers, regardless of the success of their safety records, are now subject to workplace inspections.

All scheduled OSHA inspections will now cover four principal areas:

1. OSHA will review the employer's compliance with the Hazard Communication Standard.
2. OSHA will conduct a review of the employer's OSHA-required record keeping.
3. OSHA will conduct at least a brief site inspection
4. OSHA will examine personal protective equipment in use.

Preinspection Conference

Before the start of an inspection, the compliance officer is required to explain the purpose and scope of that inspection.

This preinspection conference provides the employer with a general understanding of the coverage of this OSHA inspection.

It also provides the company's on-site supervision with the opportunity to ask OSHA for additional information, notify the employer's home office about the inspection, and receive company instructions.

Companies should always question the inspector regarding the following areas: the purpose of the OSHA inspection, how your company was chosen for the inspection, the qualifications of each inspector, and what the inspection will cover, if the company, in fact, allows it.

Conversations with OSHA inspectors should be recorded, if permitted by all parties involved. The employer should note the date, time, circumstances, and persons involved in any conversations.

An OSHA inspector on your premises without a warrant is a guest. You have the right to limit or restrict the inspection only to the areas relevant to the purpose of the inspection.

For example, if the OSHA inspector is on site in response to an employee complaint of inadequate fall guarding on a specific building or floor, you may restrict the inspection to only that building or that floor.

If the OSHA inspector accepts the limitations you place upon the inspection (remember to always detail these limitations in writing), he/she must honor that agreement. The Supreme Court in the Barlow Decision made the following statement:

> "(An OSHA inspector) without a warrant stands in no better position than a member of the public."[2]

Keep in mind that if OSHA inspectors can find anything, in their opinions, to cite as an apparent violation, they will cite it. Their opinions are not fact, and their understandings of the realities of construction operations and your construction site may be woefully inadequate.

OSHA Credentials

An employer has the right to know who is requesting access to the work site(s). OSHA compliance officers must present their credentials to the employer or the employer's agent when arriving on site.

You should inspect the credentials of each person involved in the inspection process and record their names and other identity

[2] Marshall v. Barlow, Inc. (436 U.S. 307, 1978).

information. You can also question the compliance officers about their training and experience in construction operations.

Knowing the limitations of the inspector's construction experience can be valuable later in contesting citations based on their opinions.

Since OSHA inspectors must present their credentials to the highest-ranking company agent on site, site personnel may request that OSHA delay its inspection for a reasonable time until that company agent is available. However, if OSHA inspectors will not delay the inspection, the next highest-ranking manager on the job must deal with the inspector.

Under the OSHA Act, only representatives of the Secretary of Labor may conduct inspections. Employer representatives and recognized agents of the employer's workers have the right to participate in the inspection process.

The employer or agent has the right to question anyone without proper credentials to determine an OSHA affiliation or right to participate in the inspection. The employer has the right to refuse admission to the site to any Compliance Inspector or persons without proper OSHA credentials.

Conduct of the Inspection

The employer should find out how OSHA compliance inspectors intend to conduct their inspection. Employers should ask key questions like the following:

1. Will OSHA conduct a physical site inspection?
2. What records will OSHA review?
3. Do the inspectors have their own personal protective equipment?
4. How much time will the inspection require?
5. What equipment will be used to record this inspection?

These questions are far from idle talk. For example, the *OSHA Field Operations Manual* outlines OSHA's use of camera and tape recording equipment. Employers or workers may refuse OSHA permission to take their pictures or record their statements.

Recently, some OSHA area offices have been using videotape equipment to record inspections. OSHA's Field Operations Manual does not currently cover the use of this equipment in inspections. Therefore, there are no rules for its use and no rules to protect the employer from misuse.

Employers should not allow videotaping until OSHA has developed official guidelines for the use of this equipment.

The OSHA Act guarantees employers the right to reasonable, orderly, and fair inspections. Entry onto the work site must be at a reasonable time, within reasonable limits, and conducted in a reasonable manner.

When an employer feels any part of the inspection process is unreasonable, he or she should discuss these concerns with the compliance officer. Employers have the option of restricting any portion of the OSHA inspection they feel is unreasonable. Remember, if the OSHA inspector will not agree to restrict the scope of the inspection in writing, you have the right to refuse permission to inspect.

Superintendents can request that OSHA delay all or part of the inspection until they have had time to consult with their company's management.

The employer has the right to avoid undue disruptions of the work schedule. If a critical operation is underway that the inspection would disrupt, the employer has the right to request that OSHA delay that part of the inspection until a later time or date. For example, the employer can restrict OSHA's access to a critical concrete pour area to prevent any disruption of that construction operation.

Remember, in an OSHA inspection, the mission of OSHA is to cite you, while your mission is to prevent citations. When you realize that you and OSHA are adversaries and act accordingly, you are in the best position to protect yourself.

Employer Conduct During OSHA Inspections

Discussion by management with OSHA regarding its concerns about the inspection are important. Those discussions set the groundwork for the employer to contest the inspection later. Detailed notes or recordings on those discussions should be maintained.

By conducting themselves properly during the inspection, on-site supervisors protect the rights of the employer and develop the basics for defense of alleged citations.

To protect the employer, a representative of that employer

should accompany each compliance officer during a work-site inspection—that means being constantly vigilant.

There is a key rule for employer representatives accompanying compliance officers during on-site inspections. Never volunteer information to the OSHA inspector, even if you think it is favorable to you. This is one case where anything you say can and will be used against your employer. Polite silence is not only golden, but it is necessary. There are no requirements for the employer or his representatives to answer any questions posed by OSHA.

There exists a normal human tendency to respond to questions. The employer gains nothing by conducting an oral defense of an alleged citation with the compliance officer on site.

Instruct the employer representatives not to make judgmental statements or agree with any accusatory statements made by the OSHA compliance officer.

When questioned, you as a representative of the employer are not required to answer. You may also say "no comment." You can even ask the OSHA inspector to give you clear assurances in writing that none of your answers will form the basis or any part of a citation.

OSHA's request for information is, in essence, an attempt to use your answers to convict your employer of an OSHA citation. That request is hardly fair.

The sad fact is that most employers do OSHA's job for them. They convict themselves of OSHA violations by giving statements that support the alleged citation during the inspection process.

Your primary job, when accompanying the OSHA inspector, is to begin to build your company's defense regarding citations.

The OSHA Act, under Section 8, says that OSHA has the right to speak to employees in private. The Act does *not* provide any authority for such questioning to take place on your work site while employees are on your payroll.

Employers can advise the OSHA representative that employees are being paid to work. OSHA can, therefore, arrange an interview with an employee at a place and time of his or her choosing. However, you as the employer have no obligation to provide such a place and time during working hours, and you should not.

The compliance officer has the right to examine machinery or equipment in the workplace. The inspector can also take pictures and samples and employ other reasonable techniques to complete the investigation process. You, however, should refuse to provide any assistance, such as conducting a demonstration of equipment for an OSHA inspector.

The employer representative is the only eyes and ears the company will have during the inspection process. That person, therefore, should take detailed notes during the inspection for the company to use later, if necessary.

These notes should identify the areas, equipment, and machinery inspected, along with any employees and other personnel in the area interview.

Whenever the compliance officer takes a picture or sample, the employee representative should do the same. It is a good practice for the employer representative to take more than one picture or sample from different perspectives.

The employer representative should make a full written report after completion of the inspection. That report should include labeled samples, photographs, and any other "evidence."

The employer representative assigned to accompany the compliance inspector should thoroughly understand the work in progress and the applicable safety regulations.

He or she should also have basic equipment, such as a notebook, camera, and any necessary sampling gear or personal protective equipment.

The OSHA inspector is seeking to establish the existence of noncompliance with an OSHA standard. To do that, he or she must show the employer's knowledge that such a violation exists and that employees are exposed to that hazard. OSHA must also show that a means and method for compliance with the regulation exists. Without these three elements, OSHA will have a hard time proving any violation.

The OSHA inspector will use what she or he sees, hears, and believes to justify citations against the employer.

Employee Representatives

Employees represented by collective bargaining agreements have the right to have an employee representative accompany

OSHA inspectors during a physical inspection of the workplace. These employee representatives do not have the right to review confidential records or participate in the record-keeping phase of the inspection, and they also do not have the right to actively assist OSHA in the inspection process.

When no authorized employee representative is available, the Act provides that the compliance officers consult with a reasonable number of employees regarding job-site safety. OSHA compliance officers are instructed to question employees regarding overall safety and health knowledge.

Remember, however, that you do not have to provide the facilities for such interviews or to allow workers on your payroll work time to meet privately with OSHA.

Imminent Danger

If the compliance officer concludes that a condition or practice exists that could reasonably cause death or serious injury, that officer must take action. The inspector must inform the employer or his representative of his findings and attempt to get the employer to improve the condition. If an employer refuses to abate the condition, the compliance officer must personally inform the affected employees of the danger.

The compliance officer has no authority to shut down the job site or mandate abatement of a condition.

If the employer does not agree with OSHA's assessment of an imminent danger situation, OSHA can only seek a civil injunction to restrain or remove employees from imminent danger conditions.

Employers have the right to discuss their concerns and disagreements with the compliance officer. They can state their reasons why they believe the condition does not present a serious risk to employees.

Employers should develop the habit of routinely questioning all OSHA's actions and assessments during inspections. The OSHA inspector assigned to your work site may not have the experience or education to make a valid judgment on the seriousness of a particular work hazard.

Correcting Safety Hazards

The abatement of all hazards on the job site is a good business practice and makes good common sense. Abatement, however, should always proceed in the safest manner possible. There is an impetus to correct any alleged unsafe condition in haste whenever OSHA's on the site.

The speed with which you correct a hazard is not relevant to the issuance of an OSHA citation. Often, employers, in that haste to correct an alleged violation, expose themselves to additional citations and their employees to unnecessary risk.

There are numerous instances when employers have been cited for safety violations created by their attempts to correct a reported safety hazard during an OSHA inspection. For example, one employer was cited because he sent a crew of carpenters to reinstall a guardrail that had been removed from an elevated work area.

The OSHA inspector, seeing the employees now exposed to the unguarded fall hazard, upgraded the citation from nonserious to serious.

Closing Conferences

After the completion of the inspection, the compliance officer will request a conference with the employer. The OSHA inspector will use this opportunity to look for information to use against the employer.

For example, OSHA might ask you, "Were you aware of the missing guardrail on that floor?" You can be sure that the OSHA inspector will record your answer if it indicates you knew of the existence of any hazardous condition. OSHA needs to document your knowledge of an existing hazard to prevail in a citation.

Because of the potential pitfalls, many employers are requesting that OSHA submit any questions in writing and refusing to verbally discuss specifics with the compliance officers. More and more employers are declining to attend closing conferences or remaining silent throughout the proceedings.

During closing conferences, the compliance officers will informally discuss their findings, including apparent health and safety violations uncovered during the inspection. The actual citations will be issued at a later date and will be delivered to the employer by mail or directly by an OSHA representative.

When those citations are issued, OSHA will use the existence of a closing conference to its benefit. It will indicate that OSHA clearly explained the nature of the citation during the closing conference. This makes it more difficult for the employer to claim that the citation is vaguely worded.

If you decide to attend and participate in a closing conference, use it to your advantage. Do not make any admissions that something is wrong or that you have previous knowledge of a hazard's existence.

Do not argue with the inspector over the rationality or validity of the citations. The whole OSHA citation process is unreasonable, from an employer's viewpoint. Debate will not change the inspector's mind or prevent the issuance of a citation.

Do not answer questions on what the employer considers a feasible time or method for correcting alleged violations. OSHA must set the abatement time for any citation. OSHA cannot cite you when they cannot determine what constitutes adequate abatement of a particular violation.

A discussion of feasible abatement practices with OSHA works against you. OSHA will use your comments about abatement to illustrate your knowledge of a hazardous condition or alternate method of protection. Again, you are helping OSHA to convict your company of violations when you provide their inspectors with information.

Most OSHA inspectors do not understand your work operations or schedules. If OSHA is unsure of what constitutes a feasible means of abatement for your work site, it weakens OSHA's ability to prove the citation valid in court.

Employers can use the closing conference to question OSHA about the specifics of the violation, i.e., what OSHA feels constitutes feasible abatement and abatement times. A record of the answers made by OSHA assists employers in defending against citations.

Your goal in a closing conference with OSHA is to listen and to use that conference to gather information from OSHA that will strengthen your defense of an OSHA citation.

When you engage in a debate with OSHA over the feasibility

of its suggestions, you are wasting your time. Your debate only gives OSHA more information to use against you. Telling the OSHA inspector you need one day or ten years to abate a violation admits that the violation exists. It further indicates that you were aware of a method to abate that condition.

There are no provisions in the OSHA Act for the participation of employee representatives in the closing conference. Under OSHA, the role of the employee representative in the inspection process ends with the completion of the inspection.

Contesting OSHA Citations

Employers may contest any OSHA citation and may withdraw a notice of contest at any time. Contesting an OSHA citation is nothing more than asking for a chance for a fair hearing.

Employers who receive a citation from OSHA have fifteen days to contest that citation. Contested citations then come under the jurisdiction of the Occupational Safety and Health Review Commission. When contested, a hearing will be scheduled. The review commission is independent of OSHA and has nothing to do with the inspection or citation process.

Unfortunately, OSHA's own statistics show that 94 percent of alleged citations are never contested.

Most employers feel that OSHA citations, justified or not, are simply a part of doing business. In addition, the feeling is that it costs more to fight the fine than to pay it. There are several reasons why this approach is not good business.

OSHA citations become public record. As such, access to your company's citation history is available under the Freedom of Information Act. Many employers involved in litigation find out how quickly this history comes back to haunt them.

OSHA has recently changed its policy on citations. Your company can now be cited for willful or repeat violations based on your company's overall citation history, even when those citations occur in separate company facilities or geographic locations. Under this policy, these repeat citations can cost you up to $10,000 each.[3]

[3] In March of 1991, OSHA is expected to announce a major increase in fines for a workplace violation. Existing fines may be increased up to 700 percent of their present level.

OSHA has recently enacted a policy of incident by incident citations. Under this policy, citations for each alleged violation can be multiplied by the number of occurrences or exposed employees. For example, one violation for an unsafe trench of $1000 could be multiplied by the number of employees working in that location. This type of "creative" mathematics accounts for those recent $100,000 and $1,000,000 fines issued by OSHA.

A legislative push is underway to increase the fines for OSHA citations. This legislation could also make employers and individuals in company management civilly and criminally liable for safety violations.

State and local courts are pushing to increase prosecution under criminal statutes for workplace safety and health violations. If this legislation passes, a company manager could be subject to suit and criminal violation for workplace accidents.

Some owner groups are using OSHA citations as a method to determine an employer's ability to bid on work. Your OSHA citation history could limit your company competitiveness.

OSHA compliance officers have often been criticized for not having the experience or training to conduct adequate inspections, or accurately determine the occurrence of violations. OSHA estimates that 40 percent of citations issued are considered frivolous by the employer. Yet, these citations may well affect your company's ability to compete for business and directly affect your company's pocketbook.

GROUND RULES FOR CONTESTING OSHA CITATIONS

This section will provide a basic overview of the process involved in contesting an OSHA citation. More detailed material and information are available from standard published sources and trade associations. For more information, refer to the list of standard publications and information sources contained in Chapter 13.

You do not need an attorney to contest an OSHA citation or to appear before the OSHA Review Commission. However, there are times when the use of an attorney would be a good choice.

As a rule of thumb, if the issues in the case are clear-cut and do not involve complex legal issues, you may feel comfortable in representing yourself. If you are uncomfortable in the "OSHA arena," consider retaining a qualified person to represent your interests.

Regardless of who represents you, there are several things you can do to improve your company's defenses and ability to win the case.

The first step an employer should take in contesting an OSHA citation is to request an informal hearing with the OSHA Area Office Director.

That hearing provides the employer with an opportunity to discuss the citations, to obtain an understanding of OSHA's position, and to know the penalties and abatement period proposed.

This hearing may allow you to reach a compromise suitable to both OSHA and the employer without going to court. Be aware that the citation history you generate may be more troublesome that any proposed fine. This informal conference does not serve as official notice of an employer's intent to contest any citations. If nothing else, it provides you with the opportunity to explore OSHA's case against you.

When contesting an OSHA citation, the employer must notify the OSHA area director in writing within fifteen working days from the date that the notice of a citation and penalties is received.

Send the written notice by registered mail to the OSHA Area Office Director. The address of the OSHA area office and name of the area director appear on the OSHA notice of citation.

Employers may object to any or all of the alleged violations and the proposed fines or the date for abatement of the violations.

There is no standard form or format for the employer to use in contesting OSHA citations. Two sample notices of contest are provided in Figures 4–3 and 4–4 of this chapter.

When an employee files a notice of contest, the time the company has for correcting those alleged OSHA violations automatically becomes extended. (OSHA calls that correction time the "abatement period.") Employers, however, cannot contest a citation simply to delay abatement.

Only when the employer challenges the validity of the specific OSHA violation does the abatement period become extended. OSHA-mandated abatement periods may not be prac-

Date of the letter

From: Company Name, Address, and
Telephone Number

To: OSHA Area Director
U.S. Department of Labor
Area Office Address

RE: Notice of Contest of an OSHA Citation and Notification of
Penalty, OSHA NO. (XXXXXX) Issued This (Day, Month, Year)

Sir/Madam:

This is to inform you of our intention to contest the Citation referenced above (Citation No. XXXXXX), including all alleged citations, proposed penalties, and the means and dates of abatement.

Sincerely,

(Name and Title of Company
Representative)

cc: file, Congress

Figure 4–3 Sample Employer Notice of Contest of All Parts of an OSHA Citation

tical for employers. Employers should, however, correct any unsafe conditions as soon as possible.

The area director for OSHA will forward the employer's notice of contest to the Occupational Safety and Health Review Commission (OSHRC). The review commission is not a part of

Date of the letter

From: Company Name, Address, and
Telephone Number

To: OSHA Area Director
U.S. Department of Labor
Area Office Address

RE: Notice of Contest of an OSHA Citation and Notification of Penalty, OSHA NO. (XXXXXX) Issued This (Day, Month, Year)

Sir/Madam:

We wish to contest the citation, penalty, and means and date of abatement for items number (list each item number contested) of the citation issued (Day, Month, Year).

Sincerely,

(Name and Title of Company Representative)

cc: file, Congress

Figure 4–4 Sample Employer Notice of Contest of Specific Items on an OSHA Citation

the Occupational Safety and Health Administration but is a part of the United States Department of Labor.

OSHRC will notify the employer that they have received a copy of the notice of contest and assign a docket number identifying that specific complaint and case. OSHRC will also send the employer a "Notice-to-Employee Form," advising employees of their rights to participate in the case proceedings.

Employers must post this notice at a central location and return a postcard form provided by OSHRC indicating that the employee's notification has been accomplished.

Employers should maintain a case file with all the documents pertinent to the case and must allow employees to review and copy these documents upon request.

Within 30 days after receipt of a notice of contest, the attorney for the Secretary of Labor must file a complaint with OSHA and send a copy to the employer.

The employer must then file a written answer to that complaint with OSHA within 30 days. That answer details the employer's acceptance or denial of each alleged OSHA violation and the basics of the employer's defense.

In most cases, employers will utilize the services of an outside professional versed in the OSHA citation process to file that answer. Some employer trade associations provide assistance at the local level in contesting citations and preparing documents.

An OSHRC review commission judge will be assigned to the case and a scheduled date for a hearing will be established. Employers will be informed at least 30 days prior to that hearing date of the time and place.

You must notify your workers of this hearing date. Most employers will accomplish this by posting the notice they receive in the workplace.

After the hearing is concluded, the Review Judge must file a decision with the OSHRC within 21 days. The review commission has 30 days to decide if they will review the individual judge's decision.

If the decision is not reviewed, it becomes final and binding on all parties. A copy of the judge's decision and a Notice of Decision form will be mailed to the employer. The employer has the right to seek a judicial review of that decision.

The employer need not be represented by an attorney in contesting OSHA citations. Anyone appointed by the employer may represent the employer in an informal conference or at an OSHA Review Commission Hearing.

If you, as an employer, are successful in contesting an OSHA citation, you may be eligible to collect your litigation expenses and attorney fees under the "Equal Access To Justice Act."

The Employer Edge Against OSHA

Contesting an OSHA citation is not as complicated as it may seem. It does involve time, effort, and knowledge, but it can be done and done successfully. Table 4–3 provides a simplified overview of the employer's responsibility in contesting an OSHA citation.

There are some ways to increase your chances of winning that give the employer an edge.

Start with a truthful assessment of the reasons for contesting an OSHA citation. Was the alleged citation valid? Was it reasonable? Do you feel that OSHA has made a mistake?

Perhaps, one of the biggest fears of people in contesting OSHA citations is the thought of going up against the government with all its rules and regulations—the "you can't fight City Hall" syndrome.

In reality the government, including OSHA, is simply composed of people—ordinary people, like you and me, who make mistakes and have their own views and opinions. As for those

Table 4–3 Employer Responsibilities in Contesting OSHA Citations

1. The employer must file a written notice to contest the alleged OSHA violations, penalties or abatement date within fifteen working days of the date they received the OSHA citation and notice of penalties.
2. After filing the case, the employer must certify to OSHA that employees have been notified about the case. Employees have the right to inspect and copy all documents pertaining to the case.
3. The employer must answer any complaint filed by the secretary of labor within 30 days, after the date of service, and notify employees about the hearing date.
4. The employer must attend the hearing, and the employer must abide by the hearing results.

rules and regulations, they apply to both parties, and you can actually use them against OSHA.

OSHA outlines the procedures for conducting OSHA inspections in a publication called the *OSHA Field Operations Manual* which details the steps, methods, and procedures an OSHA inspector must take during the inspection of your work site.

Remember, earlier we outlined the need for an employer representative to accompany each OSHA inspector while on site. Part of that representative's job was to observe and report on the conduct of the OSHA inspection. How OSHA went about conducting your inspection is important. If they did not do everything they were supposed to do, if they bent the rules or cut corners, they will lose at the hearing.

Before deciding to contest any OSHA citation, you should review the sections within the Field Operations Manual pertaining to the type of inspection conducted in your workplace. How does what they are required to do compare with what they actually did on your site?

As we said before, OSHA is composed of people just like you and me—people who sometimes cut corners and fail to follow all the rules and procedures.

If the multitude of OSHA rules and procedures are not properly followed, your chances of contesting the citation successfully are greatly increased.

The *OSHA Field Operations Manual* is available for a cost of under $30 from the United States Government Printing Office (GPO). You will find the address for the GPO in the reference material at the end of this book. Every employer governed by OSHA should invest in a copy of the Field Operations Manual.

We sometimes forget that OSHA is a part of the government. OSHA's boss, the Secretary of Labor, answers to Congress. OSHA's funding comes under congressional approval. OSHA is, therefore, sensitive to political pressures. Every time you write to OSHA regarding any complaint, you should indicate in your letter that a copy has been sent to your federal senator and representative.

OSHA may deny it, but mail indicating that a copy has been sent to Congress receives special attention from OSHA and quicker action and response.

Time will pass between when you file your notice contesting

the citation, and when the OSHRC hearing occurs. Employers should use this time to gather information to assist their cases.

When OSHA prepared the complaint citation they issued to you, they prepared three reports:

1. The Safety and Health or Accident Inspection Report Form (OSHA Form 1);
2. The Narrative (OSHA Form 1A); and
3. The Worksheet (OSHA Form 1B).

You should file a request (or call a motion) with OSHA to produce all documentation in your case. Figure 4–5 contains a sample request letter for production of documents from OSHA or a more formal motion can be drafted by your attorney. Always

In accordance with Rule 34 of the Federal Rules of Civil Procedures, (Company Name) requests that The Occupational Safety and Health Administration produce in their entirety the following materials related to OSHA's (Date of Inspection) inspection of (Place of Inspection) and subsequent proceedings for examination and copying:

1. All inspection reports, notes, letters and documents including OSHA's basic field report forms 1, 1-A and 1-B.
2. All written statements.
3. All recorded statements and materials including pictures, videotapes, and sound recordings.
4. All charts, diagrams, and drawings.
5. All reports and materials related to any tests, measurements, samples, or observations that were noted, taken, recorded, or analyzed.
6. Any statements, letters, reports, notes, or documents from any OSHA employee, consultant, agent, or contractor related to this proceeding.
7. All materials related to the qualifications or expertise of all OSHA employees involved in the above referenced inspection.

Figure 4–5 Sample Request for Production of Documents by OSHA

8. All materials related to the selection criteria, schedule, decision, and conduct of the above referenced inspection, including the OSHA inspection warrant and application.
9. All materials related to any written or oral communication with any past or present employee of (Company Name) related to the above inspection or proceedings.
10. The full text of all standards, rules, codes, regulations, or laws upon which any alleged noncompliance is based.
11. A complete list of all organizations, persons, associations, or agencies OSHA relies upon or uses within this inspection or proceedings to determine what is "acceptable" for employer's equipment safety or safety compliance.
12. All additional materials related to the above referenced inspection and subsequent proceedings.
13. All other materials related to any other inspection or proceeding of (Company Name) workplaces conducted prior to the (Date of Inspection) inspection and subsequent proceedings.

All the above materials are requested via mail in a legible form

within 30 days of the date of this motion.

(Employer Representative)

Figure 4–5 ***(Continued)***

include a reference to the OSHRC case or docket number in your correspondence.

OSHA has 30 days to respond to such a request, or to explain why they will not. If OSHA will not honor your motion to produce documentation, you can then make a motion with the court to compel production of such documents.

It may also benefit your case to submit questions called "interrogatories" to OSHA. You may ask questions regarding the circumstances, and basis of the citations issued against you, the documents and evidence in support of that citation, and the qualifications of the inspectors.

In essence, you are requesting OSHA to discuss the facts in your case in detail under oath, before going to court.

After receiving answers from OSHA to your interrogatories, you may wish to file a request for admissions in your case. These admissions are statements of fact that the other side in a case can respond to as either admitted or denied.

The value of admissions to you is that it pins down OSHA to the specifics and facts of the case. This prevents the real possibility of OSHA charging you with one violation, and getting you convicted of a different violation at the hearing.

For example, you can use a request for admission from OSHA that defines the hazards involved in each citation. When OSHA agrees to that stipulation, you have restricted OSHA's case only to those specific hazards.

Let's say that you stipulate the hazards that OSHA seeks to avoid are those specifically contained in the original OSHA inspection reports. When OSHA agrees to that stipulation, they are limited only to the specific issues contained in that inspection report. OSHA, therefore, cannot at the trial introduce or argue a new violation or interpretation of the regulations not stipulated in the original inspection report.

Requests for admissions are difficult to write, and you may want to obtain legal advice in preparing them.

When OSHA does not respond to your questions or requests for admission within the 30 days, they will be deemed as admitted to the court. If all this seems like a lot of bother to you, look at it from the perspective of attorneys for OSHA. They must prepare a large number of cases for hearings in front of the review commission. You have only one case to prepare.

There is little reward or incentive for an OSHA lawyer to use up a lot of time and energy for your one case.

OSHA resources are not endless. Each person in OSHA must answer to someone higher up. They must constantly justify their costs, time, and expenses.

OSHA may also submit interrogatories or requests for admissions to the employer. This is *not* the time to argue your case with OSHA. Your purpose is to state the facts and to carefully word your answers.

Be honest while trying not to hurt your case. Short, concise sentences directed to answering the specific questions posed are what you want. The idea is to give away as little as possible about your case, without providing proof of any violations or any evidence that can be used against you.

Employers should only respond to the question asked, not to what they think the question means. Again, you may want to seek legal advice to prepare your answers. You have the right to object to answering any question or to request a clearer explanation of the scope or specifics of the questions OSHA asks.

OSHA REVIEW COMMISSION HEARINGS

Contested OSHA cases are heard as near as possible to the location where the citation was issued. Although the word "hearing" sounds ominous, don't worry.

A Review Commission Judge presides over these hearings without a jury present. The hearing is conducted informally, almost like small claims proceedings. The review judge will take great pains to explain the proceedings and/or points of law to you if you are not represented by counsel.

Both parties will have the opportunity to present evidence and witnesses in the case and to cross examine the other party.

Although it is impossible to list all the rules and considerations of trial practice, some general points are important.

Testimony that is not first-hand knowledge is hearsay and not admissible. You can object to the admission of any statement or evidence not supported by facts in evidence.

For example, OSHA inspectors are often portrayed in these hearings as experts. This allows their opinions to be presented as factual evidence. You do not have to accept OSHA's portrayal of the inspector as an expert on your construction operations.

How can the OSHA inspector be considered an expert when OSHA's own Field Operations Manual does not allow compliance officers to sign written communications regarding any applications or interpretations of the OSHA Act?

In the hearing, make sure the training, experience, and credentials of the inspector as it relates to your industry, operations, and work site are fully explained.

Keep one point in mind—you know more about your company, operations, and work site than any OSHA inspector. If anyone is an expert on your business, it is you, the employer.

When opinions are asked for or conclusions are implied, relative to OSHA's expertise of your business, *object*. *Always* make OSHA stick to the facts, not the opinions, in the case. The main reasons you have taken all the trouble to obtain specific information on the violations OSHA is alleging is to force OSHA to stick to those issues alone.

A fundamental rule of evidence is that to be admissible, it must be material to the charges against you. What happened outside the scope of the specific violations for which the OSHA inspector cited your company, or the time frame of the citation period, are not relevant to your case.

OSHA must prove its charges against you. You must prove only that the defenses you have raised in your response to OSHA's allegations are valid.

The burden of proof of violations is on OSHA. If OSHA fails to prove that a violation occurred, that you knew about the violation, that your employees were endangered by that violation, or that a feasible method existed to abate that violation, it cannot win its case against you.

OSHRC statistics are in your favor. In most OSHA review cases, the employer is granted at least partial relief from the OSHA citations issued. In short, the employer wins more often than he or she loses.

Chapter 5

Selling Safety

INTRODUCTION

This chapter discusses basic safety measurement systems, safety statistics, and the benefits of accident cost analysis.

SELLING SAFETY

To be effective, safety, like any commodity, must be sold within its potential market—to both the worker on the job and to the chairman of the board. When we are not able to communicate our safety message to the field and convey our safety needs to the decision makers, we become ineffective.

Selling to Management

The safety profession is unique in that we, as safety professionals, attempt to sell ourselves from a negative standpoint. We measure the rate of our failures, not the rate of our successes.

In effect, we say, "we are doing better this year because we only had this number of people injured or killed." We concentrate on the number of losses, not on the number of wins.

We talk of doom and gloom, "You must do this because the law says so," and speak volumes of humanitarian reasons why safety is important.

We fail to speak in the universal language of business—dollars and cents. How much is this idea going to cost? How will it affect the bottom line?

Having become separated from production, profit, and cost-saving considerations, we wonder why we have a tough time selling a new safety idea to management.

Imagine, if you will, the corporate conference room at budget time: We, the managers of the widget department, are explaining to the boss why we should invest additional funding into our department because, after all, "we need widgets and we expect to lose only so much money this year anyway." Not a very effective presentation of our position, is it?

Yes, it is difficult to measure how many accidents we didn't have or how much money was saved by the accidents we didn't have. It is relatively easy to determine how much accidents cost, what costs were involved, and where improvement can be made.

Numerous cost data studies related to accidents and injuries are available. For example:

1. Accidents cost American industry 133.2 billion dollars in 1987, according to The National Safety Councils Accident Facts for 1988.
2. Lack of adequate safety costs the construction industry 8.9 billion dollars in 1980 or 6 1/2 percent of all money expended on American construction (137 billion dollars), according to a Business Round Table (BRT) Report.
3. The BRT also surveyed companies with good safety programs and found that, in 1980, they experienced workers' compensation claims of 6.1 cents per hour worked compared to the national average Workers' Compensation claims of 16.9 cents per hour.

Accident costs, overall, can be measured by direct and indirect costs.

Direct costs involve those insured costs such as:

1. Medical bills
2. Insurance premiums
3. Liability costs
4. Property loss dollar values.

Indirect costs involve those uninsured costs, such as reduced productivity, project delays, replacement costs, costs of investigation and administration of claims, court costs, and awards arising from accident cases.

A 1980 report on the average costs of safety program administration indicated that safety program costs are 2.5 percent of the direct labor costs. That safety program cost included:

- Salaries for safety, medical and clerical personnel
- Safety meetings
- Tool and equipment inspections
- Site inspections
- Safety orientations
- Personnel protective equipment
- Health monitoring, i.e., respiratory fit testing, etc.
- Supplies and equipment.

Safety performance, in fact, can be measured in many ways. There is always at least one benefit to measuring safety performance. According to a general business rule "What's measured is what gets done."

Systemic Measurements

Systemic measurements are the traditional method of quantifying safety performance by using loss ratio measurements, like frequency and severity. We often fail to remember that by measuring what we lost, we also indirectly measure what we have gained.

Organic Measurements

Organic measurements show the rate of improvement in a specific set of factors by comparisons. For example, we can compare audit or inspection reports to determine the level of loss or improvement in compliance, number of physical hazards, or other factors.

We can also use comparisons to measure accident trends or abstract factors like safety attitude or actions.

System Analysis Methods

System safety analysis and other such measurement systems seek to quantify or analyze accident events and occurrences as predictable variables within a system. As such, their frequency and impact can be predicted.

Every conceivable situation that could arise within each component of the system during its life expectancy is plotted along with its degree of effect. Models of system failures and mathematical formulas can then be developed to analyze likely failure modes and the consequences of any system failures.

System safety analysis has not seen widespread use within the construction industry.

Accident Statistics and Accident Data

Measuring how many accidents have occurred is probably the most familiar form of safety measurement. The Occupational Safety and Health Administration utilizes accident-data measurements compiled by the Bureau of Labor Statistics (BLS). This statistical method is used to develop a rate of accident occurrence per 100 full-time workers. This BLS-standardized method allows comparisons of your company's accident rate with other companies and industry groups.

Under the BLS, the same formula is used for all accident rate calculations: ($N \times C$ divided by H) where N is the number of accidents, C is 200,000 (a constant based on the work hours for 100 employees 40 hours a week for 50 weeks), and H is the total number of hours worked by the employees in the measured group.

This same formula is used to measure the Incident Rate for Total OSHA Recordable Accidents, Lost Time Cases, Medical Only Cases, and the Severity Rate for Lost Days. In the case of Severity Rate measurements, N becomes the total number of lost days.

For example, if a construction company has the accident data of 50 total accident cases, 40 medical only cases with no lost days, 10 lost time cases, 30 lost days, and total payroll hours for the year of 500,000, their accident statistics are:

Total Accidents—$N \times C$ divided by $H = 50 \times 200{,}000/500{,}000 = 20.0$.

Note: OSHA calls this the Total Case Incident Rate.

Medical Only Cases—$N \times C$ divided by $H = 40 \times 200{,}000/500{,}000 = 16.0$.

Note: OSHA calls this the Nonfatal Cases without Lost Work Days Incident Rate.

Lost Time—$N \times C$ divided by $H = 10 \times 200{,}000/500{,}000 = 4.0$.
Note: OSHA calls this the Lost Work Day Case Incident Rate.
Days Lost—$N \times C$ divided by $H = 30 \times 200{,}000/500{,}000 = 12.0$.
Note: OSHA calls this the Lost Work Day Severity Rate.

The Bureau of Labor Statistics publishes a report each year entitled *Occupational Injuries and Illnesses in the United States by Industry*. In that report, data collected from employers is used to estimate the injury and illness experience of employers in private industry. These injury rates show the relative level of injuries and illnesses in specific industries.

Rates for your company are comparable to your industry as a whole, and the subgroups within that industry are broken down by their SIC codes.

Values of Using OSHA Accident Statistics

1. The method is standardized for all industries under OSHA jurisdiction.
2. Data for calculating your company rate is contained within the OSHA 200 Log you are required to maintain.
3. BLS publishes data yearly which allows trends in accident rates to be tracked over long periods of time.
4. Simple comparisons can be made between your rates and other companies and industries.

Disadvantages of Using OSHA Accident Statistics

1. Government-published accident rates are estimates and, therefore, are subject to statistical errors. In fact, people often treat them as absolute truths instead of "best guesses."
2. Data gathered by BLS only counts the number of accidents recorded. Data is not collected or published on the causative factors involved in accidents. We end up knowing how many accidents occur, but not how or why accidents occur.
3. BLS data-collection practices exclude the collection of accident data from small employers and employers not under OSHA jurisdiction.

The National Safety Council Accident Facts

The National Safety Council collects and publishes annually a booklet titled *Accident Facts*. This publication contains more extensive accident data then the Bureau of Labor Statistics publication. Data on accidents on and off the job, accident costs, motor vehicle accidents, and fatalities are included. NSC's *Accident Facts* represents a well-thought-out and presented compendium of accident information.

Advantages of Using NSC Data

1. NSC's *Accident Facts* contains a more diverse collection of accident data than does BLS.
2. NSC's *Accident Facts* includes cost-factor data.

Disadvantages of Using NSC Data

1. Estimates of accident occurrences vary greatly with published BLS accident data and, therefore, there is no common comparison.
2. NSC Occupational Injury and Illness Data is based on submissions by members participating in an annual safety award work program. Data is not meant to be representative of industry classifications listed.
3. Data on fatalities is weighted by including an estimate of the number of work-related vehicle-accident fatalities that occur.

Measuring Injury Cost Savings

An estimate of your company's savings or losses in relation to the BLS national average accident rates can be made.

When your company's rates are lower than the BLS rates, use the formula:

BLS Rate − Company Rate = Number of Injuries Avoided

The number of injuries avoided times the average costs per injury equals your company's Accident Cost Savings in dollars.

When your company's rates exceed BLS rates, use this formula to measure the excess costs of those accidents.

Actual Company Rate − BLS Rate = Number of Cases Above the BLS National Average

Number of Company Cases Above the BLS Rate Times the Average Cost per Accident = Excessive Accident Costs

The average cost per accident can be estimated using the Accident Cost Analysis Method outlined earlier.

Accident cost savings or excessive costs can be used as an economic yardstick of your company's safety performance.

Costs of specific types of accidents can further be expressed as a percentage of your company's total accident costs by using the formula:

Specific Costs ÷ Total Costs = Percent of Costs

Knowing the costs per accident is often more important than knowing the number of times a particular accident occurs. When we look at the total direct and indirect cost of accidents, we can best determine where our accident prevention efforts would be most cost-effective.

For example, if a firm has 50 accidents involving minor injuries, say hand splinters at a cost of $5.00 per occurrence, and 5 accidents involving a more serious injury, say back-sprain cases at an average cost of $3,000.00 per case, its efforts in correcting the back-sprain problem would be more cost-effective.

In an ideal situation your safety expenditures, resources, and staff would be unlimited. Unfortunately, we all have budget constraints and limited resources to consider.

Selling Safety to Employees

Successful safety is mainly an educational effort, not an enforcement effort. All too often, the safety person's role is compared to the police traffic-enforcement role—keep people in line and keep them safe! Enforcement has its purpose in any safety program. Safety success can only be achieved by creating a cooperative safety environment—by motivating employees to participate in the safety effort and by educating employees to work in a safe manner.

Common motivational techniques can be utilized in the construction field, including positive reinforcement, incentive programs, goal setting, work facilitation, worker participation, and effective safety training.

Positive Reinforcement

Positive reinforcement involves two key elements—top management's involvement and an appeal to workers' egos and self-images.

Failing to plan only ensures that we plan to fail. You need to get started with the task of preparing for a training session as soon as possible.

Organizing the session involves you immediately in that training task and ensures you will not forget something important. Remember the "Five P's"—Prior Planning Prevents Poor Performance.

Organizing for safety training, oddly enough, begins at the end. We need to have a clear idea of what we are attempting to accomplish in order to arrive at our goals. The basic steps in organizing a training session include:

1. *Plan.* Know what you want to accomplish and how you are going to go about it. Once you know clearly what you want to achieve, you can determine what you must do to achieve it. The best way to set specific goals and objectives is to ask yourself very pointed, result-oriented questions.
   a. Whom am I teaching?
   b. What do I want to accomplish?

   It is very helpful to write your specific goals in succinct sentences before you begin to prepare the session.

   A lesson plan helps organize a session. The lesson plan should cover three primary areas.

   The beginning—Tell them what you are going to tell them and why it's important for them to know it.

   The middle—Tell them the information in a uniform, easy-to-understand manner, tying one group of ideas to another.

   The ending—Tell them what you told them in an easy-to-remember manner. Newscasters describe this as, "Tell them what you are going to tell them, tell them, tell them what you told them." This simple formula has worked for the news broadcast medium for the last 100 years.
2. *Reserve.* Make your arrangements for date, time, facilities, and equipment well in advance of your session and confirm before the session is scheduled.

A sample organization form for safety training sessions is contained in Figure 5–1. That form determines the what's, when's, and where's, the equipment needs, and even gives you a basic "things-to-do" list. Since getting started always seems to be the hardest step, using this form is a way to make it easier for yourself.

3. *Notify*. If you don't tell them where to come, chances are they are not going to get there. So, after you make your arrangements, it's a good idea to notify all the people who need to attend.
4. *Check*. Organizing training sessions can be easier if you take the time to talk to those people who could help you. Then, locate those resources that make your job easier. Since there are many resources available to you, why not use them?
5. *Prepare*. Good safety training is an audience-participation event. Be ready to respond to questions, know your material, and take the time to practice how you want to deliver that material.
6. *Focus*. All this leads to that moment when you step in front of the group and do that dreadful deed you've been planning now for weeks. If you remember to take control of yourself, relax, and listen to the audience with your ears and eyes, you will find that safety training really isn't so frightening after all.

INCENTIVE PROGRAMS

Incentive programs are awards given for performance at a specific level of achievement. Commonly in construction, incentives take the form of wage bonuses given to project management for reduced lost time accidents. Commercial incentive programs are also available to encourage good safety performance.

How Effective Are Safety Bonuses and Incentives?

The wider an incentive is distributed, the more impact it has on safety. Incentives confined to project managers do not seem to positively affect the overall safety record of a project. The most effective safety incentive programs concentrate on encouraging a

HOW TO ORGANIZE THE SESSION

TOPIC: ______________________________

DATE ____________ TIMES: ______________________

LOCATION: ______________________________

____

NUMBER OF ATTENDEES: ______________________

What Audio Visual Materials Will You Use:

____

____

____

EQUIPMENT NEEDS: (Circle each one needed)

Film Projector ______ mm	Displays
Slide Projector	Microphone
Carousels/Trays # ______	Podium
Screen	Extension Cords
VCR/Monitors ______ Format	Pointer
Overhead Projector	Lesson Plan
Blackboard	Chalk
Flip Chart	Handouts
Easel Board/pad	Sign Up Sheet
Tripods	Pens/markers

THINGS TO DO LIST:

1. Organize and develop a lesson plan.
2. Reserve the room and equipment.
3. Determine the seminar layout.
4. Notify the people.
5. Verify all arrangements before the date.
6. Check out your equipment and material.
7. Arrive early, set up, and run through.
8. Start the session on time.
9. Use evaluations to review and improve.

Figure 5–1 Training Organization Form

safety team spirit, friendly competition, and individual recognition.

Safety incentives and bonuses that target on personal monetary gains sometimes provide more of an incentive to fudge the accident records than to promote safety. It is also misleading to concentrate safety awards or recognitions strictly on lost-time accident rates.

The overall occurrence of accidents on a work site provides a better picture of that project's safety efforts.

Many projects with zero lost-time accident rates end up with excessive costs for a high number of supposedly minor but costly accidents and delayed litigation claims.

Incentive and bonus programs should ensure that a wide segment of the work force has an equal chance to receive some type of ego recognition for a job well done. The use of positive reinforcements affecting a wide group of employees is a more effective safety strategy than a bonus or award available to only a few.

Ego recognition can include many types of awards: craftperson of the month, crews with the best weekly or monthly safety records, etc. Effective safety-incentive programs appeal to the individual's natural desire to respond to positive recognition. They also build a team safety effort while increasing safety awareness.

An effective safety incentive program can be as simple as awarding "We're # 1" safety team hard-hat stickers to crew members with no accidents. You can even recognize the best safety improvement. Make sure that whatever the award is, it is followed by some kind of individual recognition. Individual recognition can be in the form of praise from the project superintendent, or a supervisor, or even a mention in the company newsletter.

Incentive programs should be set up around a central theme and slogan. Use realistic award criteria and apply them consistently. Using the same theme and slogan in safety training, banners, and posters reinforces safety training and increases interest in the incentive program.

Incentive programs do not have to cost a lot of money. In fact, high dollar awards or expensive prizes often make the participants more concerned with their own gains than with the purpose of the program.

WORKER–MANAGEMENT PARTICIPATION GROUPS

Worker participation is a widely used method to increase productivity, boost worker morale, and promote more active worker participation in the safety process. Most commonly seen in worker quality control circles, where workers meet and discuss problems and their solutions in production, these techniques can be adapted and used in safety programs.

The use of employee and management safety committees is not a new concept, but its use in the construction field has been limited. Many employers feel that the OSHA Act holds only employers accountable for workplace safety and health.

Therefore, employee participation in the safety-decision process will interfere with responsibility to provide a safe workplace. OSHA has stated, "employee involvement in decisions affecting their safety and health results in better management decisions and more effective safety protection."

In one sense, construction employers routinely engage in a form of worker–management safety participation. They encourage employees to report unsafe conditions. That use of the employee's craft and work-site knowledge can be expanded to involve employees' participation in:

1. Job safety analysis
2. Revision of job safety rules and procedures
3. New employee safety orientation and on the job safety monitoring and training
4. Emergency response teams
5. Safety committees
6. Accident and claims review committees.

Employers who encourage worker participation in the safety process and provide a forum for workers to express their concerns to top management are viewed as more safety conscious and more concerned with their employees.

Most construction employers already practice an "open door policy" related to workers' safety concerns. When employers ensure that those concerns are acted upon, they reinforce their company's commitment to safety. When employers take the time to tap worker creativity and harness worker energy in solving safety problems, they make safety a collective team effort and concern.

Employee Motivation

A department of energy study submitted surveys to over 1,000 craftpersons at 12 large construction companies. Workers were asked to identify factors that motivate or demotivate employee safety. That study found that the motivators rated highest were:

1. A good safety orientation program
2. A good safety program
3. Overtime
4. Good craft relations
5. Well-planned project
6. Solicited employee suggestions
7. Pay
8. Recognition
9. Defined goals
10. Open house and project tours

Demotivating factors, or those factors that decrease employee morale, included:

1. Disrespectful treatment
2. Lack of available material
3. Lack of accomplishment
4. Tools not available
5. Lack of recognition
6. Productivity urged but not rewarded
7. Lack of craft cooperation
8. Communications breakdown
9. Unsafe conditions
10. Ineffective utilization of personnel
11. Incompetent personnel
12. Overcrowding
13. Poor inspection program
14. Lack of participation in decision making

It is interesting to examine how many of these factors (cited as increasing or decreasing morale) are essentially matters of open or better communication between employees and management. Respect, recognition, training, concern, communication, and taking the time to listen make up the formula for better safety cooperation and performance.

People, it has been said, are most likely to pay attention and learn when the speaker broadcast is on station WII-FM, or the "What's In It For Me station."

ACCIDENT COST ESTIMATING SYSTEM

ACES, an Accident Cost Estimating System, is a revised and updated program based on a concept proposed by Stanford University back in the 1970s. Simple to use, ACES provides an easy method to obtain an immediate and reasonably accurate estimate of accident costs.

ACES uses a fundamental application of standard accounting procedures for estimating the direct and indirect costs involved in accidents. ACES is based on the development of a numerical value, called a multiplier, for specific types and severity of accident occurrence.

This multiplier times the worker's hourly wage rate and calculates costs for a specific accident or series of accidents.

The ACES multiplier chart has been recently revised by Training Consultants Resources, utilizing statistical data on the increases in medical premiums, insurance, lost wages, indirect costs, and wage rates over the last decade. This revision allows the original multiplier system to more accurately reflect current accident, injury, and illness costs.

Unlike other accounting methods, which detail claims or medical costs only, the Accident Cost Estimating System (ACES) has the following advantages:

1. ACES is easy to use for any accident situation.
2. ACES is compatible with present accounting and project-cost control systems.
3. ACES accounts for both direct and indirect accident costs.
4. It provides a simple method to show timely and realistic feedback on accident costs; insurance-loss runs, claim-report reviews, and medical-bill audits are not available until months after accidents occur.
5. The method is flexible, inexpensive, and simple to understand and apply.
6. The system is comprehensive; ACES has the ability to handle different types of accidents and degrees of accident severity in its analysis.

Using the Accident Cost Estimating System, (ACES) System

The key to using the Accident Cost Estimating System (ACES) lies within the Accident Cost Multiplier Schedule Sheet. (See Table 5–1.) This schedule sheet lists the average multiplier for specific types of accidents.

When you look at the cost analysis sheet, you will notice that the body parts injured are listed on the left-hand side of the schedule.

The types of injuries are listed across the top of the chart. The corresponding columns below the injury and across from the part injured contain the cost multipliers.

For example, a fracture of a finger lists the multipliers 40–610. The number on the left (40) is the multiplier for a medical only (no lost time) injury. The number on the right (610) is the multiplier for a lost-time injury.

Using the example cited above, let's assume that a laborer fractured his finger, but lost no time from work. To determine the potential cost value of this injury, you would ask the following questions:

1. What body part was involved? (finger)
2. What type of accident? (fracture)
3. Was this a lost-time accident? (no).

Looking on the schedule sheet, you find the value or multiplier for this type of injury is 40.

To determine the potential accident costs, use the following formula:
Number of accidents × Multiplier × Wage rate
This can also be abbreviated $N \times M \times W$.

If a laborer's hourly wage with fringe benefits on your project is $10.00, then the accident cost equals:
1 × 40 × $10.00, or $400.00. Therefore, the accident cost for this injury is $400.00.

In single-person multiple-injuries occurrences, where one person has more than one injury, the user addresses the cost of the single most expensive injury resulting from that accident.

Multiple person accidents in which two or more people are injured are treated as a series of individual accidents, by determining a separate value for each person. For example, if three workers in one accident had a broken wrist, a laceration, and a fractured finger, they would all be treated as separate costs.

All accidents of a similar type, with the same Multiplier, Craft Wage Rate, and Lost Time Status can be lumped together. In the formula ($N \times M \times W$), N then becomes the total number of similar accident cases, M the common multiplier, and W the common wage rate.

Accident Cost Estimating System (ACES) Limitations

This method of accident cost accounting estimates the direct and indirect costs involved in an accident. In developing the multiplier chart, a conservative indirect to direct cost ratio of 2 to 1 was used.

Studies have indicated that accidents may have indirect to direct cost ratios from 1 to 17. Your cost ratio may therefore be higher. This means that for every dollar of direct accident costs your company spends, there may be \$1 to \$17 of indirect costs.

The system of multipliers should, therefore, be reviewed, based on your company's direct and indirect accident costs and adjusted as necessary. To make that adjustment, you may add another multiplier into the basic formula to adjust for your cost ratios by using the following formula:

$N \times M \times W \times (+ - M2)$.

N is still the number of similar accidents.

M is the multiplier for that type of accident.

W is the hourly wage rate for the type of worker injured.

$M2$ is the value you assign to this particular type of accident based on your company's accident costs ratio.

If, for example, your ACES costs for a particular accident were \$400.00, but your actual accident cost averaged \$800.00, you would add an $M2$ multiplier to the formula of $+2$ ($400 \times 2 = 800$).

One of the primary values of ACES is that it readily adapts to provide realistic estimates of your company's accident costs and can be refined to fit your needs.

Table 5–1 ACES (Accident Cost Multiplier Sheet)

Numbers on the left side are for first aid only (no lost time) cases. The numbers on the right side are for lost-time cases. To obtain cost factors, multiply the number and type of injuries by the job labor rate, including fringe benefits. Symbols: K = 1000, NA = Not Applicable, @ = Each.

Injury Type or Body Part	Amputation	Strain Sprain Crush Smash Mash	Fracture	Cut Puncture Laceration	Burn	Bruise Abrasion	Other
Eye(s)	5.3K–28.8K	NA	NA	35–350	25–610	35–120	35–610
Head/Face	NA	NA	80–960	35–350	40–880	35–120	40–720
Neck and Shoulder	NA	40–830	175–960	35–350	40–610	35–240	35–830
Arm(s) and Elbow(s)	24.4K–28.8K	40–480	120–720	35–350	35–610	35–610	35–720
Wrist(s) & Hands	6.1K–28.8K	35–305	80–1.4K	35–350	40–610	35–380	40–720

Thumb(s) & Finger(s)	960 @	35–305	40–610	35–350	25–610	25–352	25–610
Chest & Trunk	NA	55–480	NA	35–960	40–610	35–350	35–1.1K
Rib(s)	NA	40–120	55–480	NA	40–610	40–350	35–1.1K
Back	NA	249–1.2K	Na–11.8K	35–350	40–880	40–610	40–1.2K
Hip(s)	NA	NA–415	55–1.4K	25–350	40–610	40–610	55–480
Leg(s) & Knee(s)	10.6K–33.6K	50–480	55–1.8K	35–350	40–610	35–350	35–960
Foot/Feet & Ankle(s)	6.1K–10.6K	35–305	55–1.0K	25–305	35–350	35–120	40–240
Toe(s)	830 @	35–175	25–305	35–350	40–240	25–120	35–240
Hernia Rupture	NA	NA	NA	NA	NA	NA	35–960

Special Consideration: Heart Attack = 3.5K, Hearing Loss = 1.2K, Death = 10.6K

Rates adjusted for increases in: medical, lost wage, insurance, direct costs and craft wages (1979–1988).

Information on a computerized version of ACES is available by writing to Training Consultants Resources, Inc., P.O. Box 491144, Ft. Lauderdale, Florida 33349.

Companies using this system need to take care to avoid the appearance of placing a dollar value on human suffering. An explanation to the employees of the Accident Cost Estimating System (ACES) Programs goals within the company's total safety program can prevent this perception.

Chapter 6

Training the Trainer

INTRODUCTION

According to the *Book of Lists,* stage fright is the foremost fear of most Americans. Why does the normal act of communicating and speaking in front of a group of people cause such anxiety? What can we do to control such fears?

The phone rings, the boss calls, and you have just become your company's safety expert. Oh, yes, the first class is next Tuesday! Congratulations? If you're like most of us, the first thing that runs through your mind is: "WHY ME ? I'M NOT EVEN A TRAINER !"

Your boss and the company believe that you can do this training job. That's why they gave the job to you. You are the best choice for the job. You know the work, and you know the people you work with.

Training is important. It saves lives; it prevents accidents, injuries, and pain.

Think back—how many training sessions have you sat through yourself? Do you remember thinking, "I could do better than that"? Well, here is your chance. You already know what works and what doesn't.

Don't try to be someone else, just be yourself. After all, you're talking about the one subject you know inside and out—your job.

If you think back, you will remember other times when you had to learn or to try something new. With some training and practice, you were able to accomplish the task. You did it before, so you can certainly do it now.

The number one fear of most people is a fear of public speaking. It's normal to be nervous. Just remember, you're not alone.

There are resources and help all around you, like your company, other supervisors, outside agencies, numerous "how-to" books, and safety professionals. Talk to them, and learn from them. Make use of all the resources available to you.

TRAINING KEYS

Prepare

Begin your preparation as soon as possible. The more you delay, the less you will get done. One of the best ways to fight anxiety is action.

Focus on the topic and task, and don't worry or let your imagination run wild. Being prepared takes away uncertainty.

Power

You have power over yourself; you are in control. The most powerful tool to fight stress is *breathing*. Normal, natural breathing calms the body and the mind. It increases the oxygen supply to the brain, reducing the outward signs of nervousness. When you breath normally, you have the breath to support your voice. When you sound good, you will feel better about yourself and the situation.

Positive

Be yourself and relax; you're here doing this job because you have the experience and the ability. Your company believes in you, so you should believe in yourself and try to project a positive image.

Present

You're prepared, and you can do it! Take a deep breath and let yourself go. You'll be pleased at the results.

THE CARE AND FEEDING OF STAGE FRIGHT—OR NOT ALL BEASTIES HAVE SUCH BIG SHARP TEETH AFTER ALL!

What Is Stage Fright?

Stage fright is really just another word for stress and a normal body's response to any demand placed on it. Stage fright doesn't really happen to us, but rather it is our reaction to what happens or what we think might happen to us. We really create our own stress; therefore, we really can learn to control it.

We learned about stage fright as children, perhaps when we did the wrong thing to get attention or were called upon in school when we just didn't know the answer.

It's all part of acceptance. We're afraid because we think we're going to fail and others are not going to like us. Stage fright often comes directly from the flight or fight reaction of survival.

Our bodies don't know the difference between excitement and terror any more than they know the difference between real and imagined dangers.

It's a misconception to believe that we can eliminate stage fright, any more than we can eliminate everyday fear or stress. We can, however, learn to identify, control, and channel fear to our advantage.

Stage fright is a master of disguise. It effects each of us uniquely. For some of us, it's trembling hands, knocking knees, and/or a squeaky voice. For others, it may be like the worried novice who said, "When I stand up to speak, my mind sits down."

Identifying Stage Fright

Some of the common symptoms of stage fright can be summarized as physical, cognitive, emotional, or behavioral.

Physical stage-fright symptoms can include dry mouth, throat tension, quaking voice, trembling, and shortness of breath.

Cognitive symptoms may include a lack of concentration, a blank-out, a rigid stare, and a loss of humor.

Emotional responses include tension, fear, agitation, aggressiveness, and the "speed-up" syndrome.

Behavioral symptoms may include irrationality, argumentativeness, and a substitute for relief, i.e., alcohol, drugs, and food.

Causes of Stage Fright

Obviously, we didn't intentionally set out to cause our own fears. Yet, certain attitudes and actions we express result in that fear, i.e., stage fright.

Let's examine together some of the ingredients that create the recipe for world-class stage fright. A little later, we will talk about some of the ways we can overcome the problems we create.

Unrealistic expectations of ourselves or our abilities often place us in a position where we just can't win. We're all human, and we all make mistakes. Yet, when we set goals we can't reach, we only invite ourselves to fail. This behavior often becomes a vicious cycle—the greater our expectations, the more we experience failures and the less we expect to succeed.

Just as we tend to expect too much from ourselves, we also expect too much from others. When the others don't come through as we think they should, we become angry, frustrated, and stressful.

Self-defeating statements weaken our effectiveness by undermining our own self-confidence. If we tell ourselves enough times that "I can't, I'm going to fail, I'm not able," we really shouldn't be surprised when we accomplish exactly the goals we set for ourselves.

Deep-seated fears and securities often show up most vividly when we're in the spotlight, or in front of a group. Fear of failure, criticism, rejection, or humiliation can easily haunt us. Fortunately, our most dire expectations usually never come to pass. Yet, we allow fear to override common sense.

Worrying is like a wart, ugly and useless. But, many of us spend so much time worrying, we must really enjoy it. Worry feeds on itself and grows stronger.

> "What if my throat gets dry?"; "What if the jokes don't work?"; "What if I don't get there on time?" Don't' Worry; Be Happy!!

Procrastinating, overscheduling, accepting unrealistic requests all result in the same thing—too much to do and not enough time. We allow ourselves, therefore, to be unprepared, to feel trapped, negative, and nervous.

The point of all this is very simple; it is not the circumstances, the people around us, or the opportunities given that cause fear. Fear is something we inflict upon ourselves. It is something only we can choose to do something about.

Controlling Fear

No one is immune to stage fright—from the Chairman of the Board to the laborer, from the Toastmaster to the novice, we all feel the butterflies when the spotlight approaches. Bill Brooks, a well-known author and public speaker, noted

> I accept the fact that I'm going to have butterflies in my stomach every time I speak, I only want to get them to fly in formation.

To help with achieving that formation, let's consider some fear control techniques.

Let's face it, stage fright is almost never fatal. Also, you're not alone! Almost every time you speak, someone is going to want to know how you keep so calm.

Never apologize or offer excuses about your speech. Apologies do not change your shortcomings, they only draw attention to them.

If you don't tell them you're afraid, chances are they're not going to know it. Some speakers attempt to gain sympathy through confession and end up losing the audience's respect and admiration. Guard against outward signs of fear. Get on with what you came to do. Get involved with what you are doing. You'll be surprised with how well you are perceived.

The number one sin in communication is to lose your sense of humor and take yourself too seriously. When an audience learns that you are willing to laugh at your own mistakes, the audience will feel free to laugh along with you instead of at you.

Breathing is one of the best ways to control stress. In addition, just before you are scheduled to begin, take a moment to relax by sitting comfortably, focusing your attention on one object, and simply telling yourself to relax and breathe.

Feeling good about yourself ensures good results and inspires self-confidence. Confidence comes from preparation and practice. When you are prepared for what you are going to say and how you are going to say it, you can cope with whatever problems arise. When you have practiced enough that you are comfortable, you can focus your attention on your audience and audience rapport.

Everyone is shy and basically self-conscious. All of us are timid about meeting strangers whether one at a time or a thousand. The audience is really just like you—they came to listen, to learn, and to be entertained.

It is always more comfortable to talk to friends, so why create problems for yourself by assuming that the audience is hostile? The more you think of the audience as "we" and "us," and the less you think of them as "you" and "they," the less isolated you feel.

Perhaps the single best advice to give on controlling stage fright is just to relax and be yourself. After all, the easiest role of all to play is the one you know best of all—**YOU!** The best speakers find their strongest advantages and use them to create their own styles.

ORGANIZING THE TRAINING SESSION

Introduction

Failing to plan only ensures that we plan to fail. Remember, we emphasized the need to get started with the task of preparing for our training session as soon as possible.

Organizing the session gets you right into that training task and ensures you will not forget something important. Remember the five "P's"—Prior Planning Prevents Poor Performance.

Organizing for safety training, oddly enough, begins at the end. We need to have a clear idea of what we are attempting to accomplish in order to arrive at our goals. The basic steps in organizing a training session include:

1. Plan. Know what you want to accomplish and how you are going to go about it. Once you know clearly what you want to achieve, you are in a position to determine what you must do to achieve it. The best way to set specific goals

and objectives is to ask very pointed, result-oriented questions.
—Whom am I teaching?
—What do I want them to do?
—What is important for them to remember?

It is very helpful to write your specific objective in a succinct sentence before you begin to prepare the session.

A lesson plan should contain three primary areas:

The beginning—Tell your audience what you are going to tell them and why it's important for them to know it all.

The middle—Tell them the information in a uniform, easy-to-understand manner, tying one group of ideas to another.

The ending—Tell them what you told them in an easy-to-remember manner.

2. Reserve. Make your arrangements for date, time, facilities, and equipment well in advance of your session and confirm before the session is scheduled.

In Chapter 5, Figure 5–1 presented a sample organizational form to assist in training programs. If you turn to that form, you will see that the form outlines the what's, when's and where's, the equipment needs, and even gives you a small "things-to-do" list. Since getting started always seems to be the hardest step, we've found a way to make it easier for you.

3. Notify. If you don't tell them where to come, chances are they are not going to get there. So, after you make your arrangements, it's a good idea to notify all the people you would like to attend.
4. Check. Training sessions can be tremendously easier if you take the time to talk to those people who could help you and locate those resources to make your job easier.
5. Prepare. Safety training should be considered an audience-participation event. Be ready to respond to questions, know your material, and take the time to practice how you want to deliver that material.
6. Focus. All this leads to that moment when you step in front of the group and do the dreadful deed you've been planning now for weeks. If you remember to take control of

> yourself, relax, and listen to the audience with your eyes, you will find that it really isn't so terrible after all.

Experience says, Aldous Huxley, . . . "is not what happens to a man, it is what a man does to what happens to him."[1]

A novice gloats over success and learns nothing useful from it. The more experienced person gains from each success.

[1] From Charles L. Wallis, *A Treasury of Sermon Illustrations* (Nashville: Abingdon Press, 1950); Copyright 1950 by Pierce & Smith.

Chapter 7

Crisis Management

INTRODUCTION

One of Murphy's laws states, "If anything can go wrong it will go wrong, at the worst possible moment and with the worst possible consequences."

Due to this cheerful disposition, some cynics feel Murphy is an optimist. I've always felt that the answer to the origins of Murphy's Laws is simpler. Somewhere in his illustrious career, Murphy worked as a construction manager with safety responsibilities.

In construction, crisis seems to become an inevitable part of the job. The purpose of this chapter is to assist you in planning for, organizing, handling, and surviving crises. The tables in this chapter can be incorporated into your company safety or crisis planning manual.

CRISIS MANAGEMENT IN THE CONSTRUCTION INDUSTRY

Construction is, by its very nature, a hazardous industry, a business where the potential for accidents exists at all work sites. The difference between a disaster and an accident is often times a matter of degree or luck—luck that several workers were not injured or killed or that the severity of the occurrence was not excessive. No business is immune to disaster and its catastrophic consequences. Yet, few contractors develop plans to effectively deal with disasters when they do occur.

Business decisions made in the panic of the moment can be damaging decisions, Decisions that can seriously effect your company's ability to continue doing business. Planning for disaster occurrences provides an orderly flow of action for your managers and employees, reduces losses, and helps to mitigate damages. The time to develop and to implement these plans is before disaster strikes.

WHAT CONSTITUTES A CRISIS?

Disasters are thought of as either natural or manmade. Natural disasters involve forces of nature like wind, storms, floods, etc., that create injuries or disruptions in business. Manmade disasters include occurrences like fire, explosions, structural collapse, riots, strikes, kidnapping, sabotage, or other actions that result in injury, loss of life, or disruptions of business. However, not all crisis situations arise from disasters.

A crisis includes any situation that can escalate negatively, come under close media or government scrutiny, or interfere with a company's business operations. In short, any occurrence that can damage your company's business can create bad publicity or threaten your company's bottom line.

PRECRISIS PLANING

The main priority of all emergency planning is to preserve human life. The coordination of the multitude of details of a control or rescue and recovery operation is the way to accomplish that goal. Coordination occurs through the application of a well-organized disaster plan that considers short- and long-term needs and actions along with the interplay of outside organization with your company.

Disaster preparation is perhaps one of the last considerations undertaken in construction planning. Most companies adopt the "it's not going to happen to me" attitude. Trusting to luck may be an easy solution, but the results when luck runs out can be frightening.

Without a plan, confusion and panic control the situation. Without a plan, lives are jeopardized needlessly. Without a plan, your company can find its reputation and ability to conduct business threatened.

Precrisis planning originates at the corporate level, but is executed at the site level. To control a crisis, flexible and realistic plans are needed. These plans should provide clear overall guidance, yet are plans that are practical and easily tailored to the individual site needs.

That planning goal is not as difficult as it may sound. For a larger company or one with multiple work sites, the framework of a disaster plan is developed at the corporate level to determine the overall flow of actions and the results desired. That overall plan must then be tailored to fit the specifics of each individual job by including specific job information. For example, for each site: who to notify, what phone numbers to use, what actions to take, and in what sequence.

THE CRISIS MANAGEMENT TEAM

The first question to answer in developing a company plan to deal with disasters is, "do I have the time and ability to prepare this comprehensive crisis program in-house?" If the answer is no, then perhaps you should consider the services of a crisis management consultant. These consultant companies can provide an audit of your company's disaster preparedness. They can save you time and money by providing packaged programs or changing existing programs to fit your company's needs.

Crisis planning starts with the creation of a crisis team. The crisis management team will vary with the size and complexity of the individual construction company. At a minimum, the team should consist of two elements—a company crisis group and a field crisis group. The company crisis group will develop and coordinate the overall crisis planning and develop the company crisis program. The field group will tailor that company plan to fit their site needs, assess the severity of crises on site, execute the initial crisis plan, and notify the company group of any occurrence.

The basis of crisis planning is the same for large or small companies—prepare a written crisis program that spells out what to do step by step.

In this chapter, we will start with a large company crisis management plan that can be expanded or contracted to fit your company's needs.

The development and execution of crisis planing starts with the appointment of a crisis manager to coordinate the effort. Ideally, this key member of the crisis team should be someone in the company top management. The crisis manager coordinates the initial planning phase and will be the key person notified when a crisis situation occurs.

The crisis team membership can be expanded to include representatives of your company's safety manager, legal council, equipment management, transportation, and media-relations department.

A field crisis team should be designated at the project level. That team will consist of the project manager, engineer, safety supervisor, and designated site personnel.

EMERGENCY RESPONSE TEAM

A chain of command is necessary to ensure management control in a crisis and the orderly execution of emergency procedures. Responsible persons on each site are chosen to coordinate that site's emergency response efforts. At a minimum the crisis team should consist of the following:

The Emergency Response Team Leader is usually the highest ranking company official on the project site. His or her duties involve:

1. Assessing the emergency situation and authorizing initiation of emergency procedures.
2. Directing all efforts to evacuate personnel, minimize loss, and contain and control the site situation.
3. Overseeing the coordination of the project emergency response effort, equipment, personnel, and outside emergency response groups.
4. Directing any necessary shutdown and rescue operations.
5. Assessing the company's home office of the extent of the emergency and response or rescue operations.
6. Assigning any additional personnel necessary to assist in the emergency response effort.

The Communications Officer, appointed by the Emergency Team Coordinator, is responsible for:

1. Controlling the use and flow of communications between the crisis coordination center and emergency response personnel and outside sources.
2. Notifying, as directed, the necessary site personnel and community response agencies.
3. Coordinating the employee role call information received from crew supervisors. The role call assures that all employees are accounted for after an evacuation.
4. Handling initial media inquires as directed by the team coordinator.
5. Notifying, as required by law, federal, state, and local agencies regarding the crisis.

Emergency Team Coordinators, appointed by the emergency team leader, are responsible for the field coordination of on site and outside emergency response personnel and response efforts. This may include on site and external , fire, medical, rescue, security, and operations personnel. The emergency team coordinator also:

1. Assumes field control over work crews and areas or operations, as directed by the team leader.
2. Reports directly to the team leader regarding, the on-scene situation and assessment, conditions, operations, facility, and personnel needs.

Crew Supervisors are on site supervisors who will be responsible for the following:

1. Coordinate their crew evacuation by established routes and procedures.
2. Follow established emergency procedures regarding area security and equipment shut down.
3. Assemble their crew at their designated safe assembly points following evacuation, and report employee role call information to the crisis coordination center.
4. Assist as directed in emergency and rescue operations.

CRISIS PLANNING PRIORITIES

The first priority in any crisis is the protection of human life. See Table 7–1. This includes the prevention of further injuries, coordination of the rescue operation, establishing site security, and

Table 7–1 Crisis Management Priorities

1. Protect human life
2. Prevent further injury
3. Contain the situation
4. Access the damage
5. Control the damage
6. Control business problems
7. Deal with the media
8. Prepare for litigation

providing protection of workers, responding personnel, and the public.

Once necessary medical and rescue operations start, actions should be taken to control access to the crisis area and the work site. Maintaining site access control will assist in preventing further injury and getting additional equipment and personnel to the crisis site as fast as possible.

It also prevents unauthorized personnel from entering into the danger area or interfering with crisis control operations.

An accurate assessment of the extent of crisis situation by on site personnel determines what actions are necessary and provides a picture of just what you are dealing with.

Notification and information must be provided as necessary to site personnel and community response agencies.

After initial response and assessment, a brief oral report should be used to alert the company crisis manager and the company crisis team.

Crew supervisors should account for their crew members and be prepared to assist in control operations as directed.

An established communications procedure between the field operations and the field crisis team reduces confusion and provides valuable information for crisis coordination.

Additional communication procedures are needed to deal with the media and information calls from families and friends. Strict controls should be used over who addresses the media and what information is released. (See Table 7–2.)

Table 7–2 Dealing with the Press

When the media or press call for an interview, follow the following guidelines.

1. Let the media know who will be the company spokes person.
2. Do not make "off-the-record" comments. If you do not want it used, don't say it.
3. Use clear understandable language. Stick to the facts, do not speculate. Make sure your information is accurate.
4. Present a positive image. Tell them what has been accomplished. Do not make predictions about the future.
5. Condense information whenever possible to emphasize key points. When mistakes are made, make corrections where necessary.
6. It is acceptable to tell the media that you do not have that answer at this time.
7. Do not say, "No Comment"; provide the information that you want communicated.
8. Avoid placing the blame on anyone or accepting blame for yourself or the company.

THE SITE-SPECIFIC PLAN

A corporate crisis plan cannot function unless that plan has been established in writing, is tailored for the specific construction project needs, and is distributed and understood by management and workers.

A specific crisis plan maintained on each site gives step-by-step direction to the project management on handling a crisis situation. Table 7–3 outlines elements of an emergency action plan.

Making a corporate crisis plan site specific is not as difficult as it may sound. Most of the material needed to tailor a corporate crisis plan for the site level already exists on the site. It just has not yet been collected and incorporated into an orderly format. This needed information includes:

1. *Site medical services.* Perhaps your company has on-site medical services, an emergency medical technician, nurse,

Table 7–3 Elements of an Emergency Action Plan

At a minimum the project emergency action plans must include the following elements:

1. Emergency evacuation plans with escape procedures and escape routes outlined.
2. Procedures for assigned employees who remain to assist in the evacuation and the control of critical equipment and operations.
3. Procedures to account for the whereabouts of all employees after evacuation. Each crew supervisor will be responsible for the evacuation of their crew to designated locations. A supervisor will conduct a check for missing personnel and provide a role call report to the designated emergency coordinator.
4. Outline of the rescue and medical duties of those employees assigned to perform them.
5. An emergency notification contact list with names and phone numbers.

or first-aid person. Maybe you use local emergency services. You all have the emergency numbers for fire, ambulance, or police posted by the phone.

Expand upon this basic phone list by including the emergency phone numbers and contact information for the crisis team on the first page of the crisis manual. Also, include an accurate description of the geographic location for the project with references to streets, intersections, landmarks, distances, and directions. This way, anyone who has to notify emergency response units can relate precise information to assist in locating the work site.

It is a good policy to invite local rescue agency representatives to visit your work site. You can use this opportunity to discuss site-specific information with them and provide orientation tours. On large projects it is a good idea to arrange for periodic on site disaster drills with community rescue and response services.

2. *Company emergency notification sheet.* List the phone numbers and sequence of people to call in an emergency situation.

You also need to provide clear directions on the order of people to call, when to call, what information to relay, and what alternative personnel to call. Be sure to include information on the procedures to be followed when notifying regulatory authorities like OSHA, EPA, and state agencies.

3. *Site inventory.* Maintain inventory sheets for equipment, hazardous materials, and a personnel roster on each job site. The equipment sheet should list the materials and heavy equipment on site that could be useful in a disaster, rescue, or crisis situation.

   Maintain a list of off site contact numbers including other company locations and equipment rental service for emergency use.

   A hazard list should indicate what hazardous materials are maintained on site along with the quantity in storage and all storage locations.

   The Hazard Communication Standard already requires employers to maintain a hazardous chemical inventory. This inventory can serve as the base list, with additional hazards like underground storage tanks, explosive storage areas, or pressurized container locations added. A list of site personnel can be as detailed as the last payroll records or simply a day-to-day head count. The name of each area supervisor and work crew head should be included.

   Such lists provides immediate on-hand information when a crisis arises, vital information necessary in an emergency stress situation.

It is a good idea on larger sites to maintain a site diagram or set of site plans to assist in crisis situation planning and response.

4. *The corporate crisis plan.* A copy of the corporate crisis plan or a detailed step-by-step procedure for dealing with a crisis situation should be a part of the site crisis response manual. A step-by-step plan is important because it helps to turn even the panicky amateur into a capable crisis

manager. The less that is left to chance, the better the possibility that vital steps will not be overlooked or forgotten when a crisis occurs.

5. *Site assessment sheet.* The necessary information for the assessment report in a crisis situation should be collected in a "fill-in-the-blank" type format. The assessment report for the corporate crisis management team should give the overall site picture. This report can only come from the field, from the people dealing with the initial crisis situation. You know what information your business needs to judge the severity of the crisis. Add that to your company's specific information to the sample crisis report form in Figure 7–1.
6. *Record-keeping information.* Any crisis has the potential to become a legal problem for your business. Your company's best defense is to establish a detailed investigative record of the crisis for later use. In today's litigious society each crisis is a potential lawsuit.

   Take the time to make and document a proper accident investigation as soon as it is safe to do so. Maintenance of these records from day one assists in a later crisis performance review. That review will help improve and refine the overall company's crisis plan. It also places your company in a better position to defend against litigation at a later date.

Preserve all accident reports, witness statements, and first report of injuries related to the crisis, along with evidence, site-access logs, and environmental, exposure, and medical records. The site-access log should note the name and contact information for all employees and visitors participating in the crisis situation. This includes emergency response units, media responders, and regulatory agency personnel.

SUBCONTRACTORS AND OTHER EMPLOYERS

If you are a prime contractor, you should provide a copy of the Company Site Crisis Manual, or notification of your company crisis policy, to each subcontractor.

You should required each subcontractor to participate in and adhere to your crisis program as a condition of employment. Each subcontractor should also provide you with emergency contact

TYPE OF REPORT: INITIAL ________ UPDATE ________
DATE OF INCIDENT ________, 19________
TYPE OF OCCURRENCE ____
PLACE OF OCCURRENCE ____
WERE INJURIES INVOLVED YES ________ NO ________
IF YES, GIVE DETAILS ON NUMBER AND DISPOSITION OF INJURIES. ____
____
____
IF NO, WAS THERE PROPERTY DAMAGE. DESCRIBE. ____________________
____
____
LIST ANY HAZARDOUS CONDITIONS INVOLVED WITH THIS OCCURRENCE:

	(HAZARD)	(STATUS)
1.	____________	____________
2.	____________	____________
3.	____________	____________
4.	____________	____________

LIST EMERGENCY RESPONSE UNITS BY TYPE AND STATUS TO DATE.

	(TYPE)	(STATUS)
1.	____________	____________
2.	____________	____________
3.	____________	____________
4.	____________	____________
5.	____________	____________

LIST EMERGENCY RESPONSE TEAM ACTIONS BY TYPE AND STATUS TO DATE.

	(TYPE)	(STATUS)
1.	____________	____________
2.	____________	____________
3.	____________	____________
4.	____________	____________
5.	____________	____________

HAS THERE BEEN REGULATORY AGENCY RESPONSE? ____________________
____
HAS THERE BEEN MEDIA RESPONSE? ____

Figure 7–1 Sample Crisis Report Form

information to include in your site manual. Other employers on multiemployer work sites should be advised about your company program and invited to participate.

ASSESSING THE SITE LEVEL OF CRISIS PREPAREDNESS

To translate the corporate crisis management written program into an action plan that will work at the site level, you need to assess the resources on hand at each site.

No one plan written outside of the work site will be viable for each specific site.

Construction sites differ with the project, construction phases, geographic location, and a host of other factors. Corporate planners will need to consider and work with the persons responsible for on site supervision. Each site should not reinvent the wheel, but simply fine tune the corporate program to fit the realities of the situation.

Each construction site needs to be analyzed to determine the resources that exist at the site level and in the surrounding community. A site crisis preparedness assessment form will assist in that analysis. (See Figure 7–2.)

1. HAS A COPY OF THE CRISIS PLAN BEEN DISTRIBUTED TO ALL SUPERVISORS AND EMPLOYEES? YES ________ NO ________
   HAS IT BEEN POSTED IN THE WORKSITE?
2. HAS A COPY OF THE CRISIS PLAN BEEN DISTRIBUTED TO:
   FIRE DEPT: YES ________ NO ________ CONTACT PERSON ________
   POLICE: YES ________ NO ________ CONTACT PERSON ________
   MEDICAL SERVICE PROVIDERS: YES ________ NO ________ CONTACT ________
   SUBCONTRACTORS: YES ________ NO ________
   CONTACT ________
   OTHERS: ________________________________
3. HAS AN EMERGENCY EVACUATION PLAN BEEN DEVELOPED FOR THE WORK SITE INCLUDING:
   a. EGRESS ROUTES FOR ALL CREWS? ________
   b. ASSEMBLY POINTS FOR ALL CREWS? ________

Figure 7–2 Site Crisis Preparedness Assessment Form

c. INVENTORY LIST FOR HAZARDOUS CHEMICALS? ________
d. EMERGENCY PHONE NUMBER AND CONTACT LIST? ________
e. INVENTORY OF EMERGENCY EQUIPMENT WITH LOCATIONS ON SITE? ________
f. SITE PLANS INCLUDING: STRUCTURES, UTILITIES, EVACUATION ROUTES, EMERGENCY SUPPLIES & EQUIPMENT. ________
g. LIST OF THE CRISIS TEAM PERSONNEL? ________
h. LIST OF ALL BUSINESSES NEAR THE WORKSITE? ________
i. LIST OF ALL PERSONAL PROTECTIVE EQUIPMENT & SUPPLIES WITH THEIR LOCATIONS? ________

4. HAVE EMERGENCY ACTION GUIDELINES BEEN PREPARED FOR SUPERVISORS AND EMPLOYEES REGARDING:

ACCIDENTS ________	FIRE ________
EXPLOSION ________	BOMB THREAT ________
SEVERE WEATHER ________	STRUCTURAL COLLAPSE _
NATURAL DISASTER ________	CHEMICAL SPILLS ________
RESCUE OPERATIONS ________	FIRST AID ________

5. IS THERE A SITE FIRE BRIGADE? YES ________ NO ________

6. IS THERE EMERGENCY MEDIAL PERSONNEL OR AN ADEQUATE NUMBER OF CERTIFIED FIRST-AID PERSONNEL ON SITE? YES ________ NO ________

7. HAVE ALL EMPLOYEES BEEN INFORMED AND TRAINED IN EMERGENCY PROCEDURES IN EVACUATIONS? YES ________ NO ________

8. HAVE ALL OUTSIDE EMERGENCY RESPONSE GROUPS (FIRE, POLICE, AMBULANCE, RESCUE, OTHERS) ATTENDED ON SITE ORIENTATION AND ENGAGED IN RESPONSE DRILL TRAINING? YES ________ NO ________

9. ARE EMERGENCY COMMUNICATIONS EQUIPMENT AND PROCEDURES IN PLACE AND OPERATIONAL? YES ________ NO ________. ARE THEY TESTED PERIODICALLY? YES ________ NO ________

10. HAS THE CRISIS MANAGEMENT TEAM BEEN DEVELOPED, BRIEFED, AND TRAINED IN HANDLING EMERGENCY PROCEDURES? YES ________ NO ________

11. HAS THE FOLLOWING PAPERWORK BEEN ASSEMBLE AND IS IT REVIEW PERIODICALLY?

Figure 7–2 ***(Continued)***

a. SITE CRISIS PREPAREDNESS FORM? YES ________
NO ________
b. EMERGENCY INVENTORY INSPECTION REPORTS?
YES ________ NO ________
c. CRISIS REPORT FORMS? YES ________ NO________
d. TRAINING RECORDS DOCUMENTATION? YES ________
NO ________
e. ACCIDENT REPORTS? YES ________ NO ________
f. ASSESSMENTS OF EMERGENCY RESPONSE DRILLS?
YES ________ NO ________

12. ARE EMERGENCY PROCEDURES AND GUIDELINES REVIEW PERIODICALLY? YES ________ NO ________

13. IS THIS REVIEW:
a. ANNUALLY ________
b. AS SITE CONDITIONS CHANGED ________
c. AFTER EACH EMERGENCY DRILL ________
d. AFTER EACH INCIDENT ________

Figure 7–2 (*Continued*)

Carefully consider the types of community services available. For example: Does this particular site have a contract with a local hospital or emergency-room facility for medical treatment? Are there trauma centers within the site's geographic location? Are there special work-related situations that would affect medical treatment like compressed air work?

SAMPLE CRISIS MANUAL MATERIALS

As we mentioned earlier, it is foolish to reinvent the wheel. Therefore the authors have provided some material to serve as a framework for your consideration.

To assist construction companies in the development of a company and/or site-crisis plan, we have provided the following information.

Tables 7–4 to 7–11 covers elements of crisis planning or guidelines for a specific type of crisis situation. These materials can be reviewed and revised to fit your company's specific needs and incorporated into your company's crisis manual.

Table 7–4 Key Points for Handling Emergencies

The following guidelines will assist project supervisors in the proper procedures for handling emergency situations.

General Guidelines

1. The company's primary concern is the preservation of life and the prevention of injury. No unnecessary risk should be taken by personnel responding to emergency situations.
2. All emergency procedures outlined in the Company Safety Program Manual are given as guidelines. Where specific procedures have been established, relative judgment should be used to determine the best course to follow.
3. No unauthorized or nonessential personnel are allowed access to the site. All visitors must report to the project office and be escorted at all times while on the work site.
4. Secure all accident and emergency sites to prevent unauthorized access or the possibility of further risk to workers or the public.
5. All emergencies will be handled by the highest-ranking company representative at the site.
6. Conduct physical inspections after the emergency situation is under control to ensure that it is safe for work operations to continue.
7. Project personnel are directed as a condition of employment not to discuss or give statements to anyone regarding emergencies without the consent of the project management.
8. All employees are required to assist as directed in any emergency operation.
9. All employees are required to conduct such assistance in the safest possible manner, using all required personal protective equipment.

Table 7–5 Guidelines for Accidents Involving Death or Serious Injuries

1. Provide necessary first aid.
2. Notify site medical or community response personnel.
3. Designate a crew member to assist emergency response personnel in locating the accident site.
4. Control the accident site by removing non-essential personnel and securing the accident site.
5. Assist in medical or rescue operations as directed.
6. Refer all outside inquires to the project manager.
7. Allow no unauthorized access to the accident site.
8. Conduct a preliminary investigation to ascertain the basic accident facts, witnesses and evidence.
9. Prepare a accident report for submission to project management by the next business day.
10. Do not give statements regarding the accident or discuss the occurrence without consulting project management.
11. Inspect the accident site to identify and abate any unsafe conditions before work commences.

Table 7–6 Guidelines for Property Damage Accidents

1. Notify the project safety supervisor or project management.
2. Protect against further damage whenever possible.
3. Whenever the possibility of injury, fire, or explosion exist, insure the safety of employees in the danger area.
4. Secure the accident site to prevent unauthorized access.
5. Conduct a preliminary accident investigation, to ascertain the basic accident facts, witnesses and evidence.
6. Prepare an accident report for submission to project management by the next business day.

Table 7–6 (*Continued*)

7. Inspect the accident site to document and abate any unsafe conditions before work commences.

8. Do not give any statements regarding the accident or discuss the occurrence without consulting project management.

Table 7–7 Guidelines for Fire Protection Planning

The Occupational Safety and Health Act requires all work projects and facilities to institute a fire protection program.

Project management will be responsible to insure that a fire protection program is instituted and maintained.

All materials and equipment should be stored, handled and maintained with consideration for their fire hazard characteristics.

When large quantities of combustible, flammable and hazardous materials must be stored for project use they must be secured. Separate hazardous material from work, facilities and other storage areas.

Make adequate arrangements on each project for on site fire protection services or coordinated fire response with local community services. A fire protection emergency plan should be developed for the site specifying:

- Evacuation procedures
- Fire response procedures
- The location of and identification of storage for hazardous materials.
- Copies of project Material Safety Data Sheets
- Emergency notification contact phone list

FIRE POLICY GUIDELINES

Consider the following guidelines when developing a project fire protection policy.

1. Integrate the fire protection program into all phases of the construction work.

2. Conspicuously locate all fire fighting equipment. Insure that it is readily accessible at all times. Inspect all equipment on a regular schedule and maintain this equipment in working condition. Immediately replace all equipment that is used or found defective.
   a. Fire extinguishers shall be rated 2A or better.

Table 7–7 ***(Continued)***

b. An extinguishers or equivalent protection will be provided for every 3000 square feet of protected building area.
c. Travel distance from any point of a protected area to the nearest extinguisher or equivalent protection will not exceed 100 feet.
d. In multistory protected buildings, one or more extinguishers or equivalent protection will be provided per floor.
e. A fire extinguisher rated 10B or better will be provided in any area where five or more gallons or flammable or combustibles liquids are stored.

3. Where sprinkler systems are installed, they will be installed and made operational as soon as construction process and applicable law permits.

4. Smoke alarms will be installed in office, storage, maintenance and facility buildings.

FIRE PROTECTION GUIDELINES

1. Where safely possible, attempt to extinguish the fire with proper equipment.
2. Notify project management immediately regarding: the fire location, status, extent and the need for emergency response.
3. Designate a crew member to meet and direct fire response personnel to the fire site.
4. Evacuate all nonessential personnel from the fire area.
5. Safely secure the fire area to prevent unauthorized access.
6. Assist as directed with fire efforts.
7. Conduct a preliminary investigation to ascertain the basis accident facts, witnesses and evidence.
8. Prepare an accident report for submission to project management by the next business day.
9. Inspect the fire site to identify and abate any unsafe condition before work commences.
10. Do not give any statements regarding the fire or discuss the occurrence without consulting project management.

Table 7–8 Guidelines for Flammable and Combustible Storage

1. Flammable and combustible liquids will be stored and transported in approved and properly labeled containers.
2. When 25 or more gallons of flammable or combustible liquids are stored, approved storage cabinets or tanks must be used. All applicable fire protection regulations must be complied with.
3. Smoking, hot work or open flames are prohibited in flammable and combustible storage or transfer areas.
4. Storage of flammable or combustible liquids outside of buildings shall be separated from all structures by a minimum of 20 feet. No more than 1,100 gallons will be stored in one area. Storage areas will be diked to contained spills and graded to divert spills away from structures and underground work areas.
5. Flammable and combustible storage areas will be posted as no smoking areas. These storage areas will be fenced in to prevent unauthorized access. All fire protection requirements will be maintained.
6. All tanks, drums and containers in flammable or combustible transfer areas will be bonded and grounded.
7. All inside storage areas for flammable and combustibles will meet or exceed the fire resistance and protection requirements. No more than 60 gallons of flammable and 120 gallons of combustible liquids may be stored in an indoor approved storage cabinet.
8. Work areas will be kept free from debris and material that could constitute a fire hazard. Work area will be cleaned as necessary to remove accumulated materials and trash.
9. All trash and scrap will be properly stored and or disposed of, in accordance with all applicable federal, state and local law.

Table 7–9 Guidelines for Explosions

1. All nonessential personnel should be evacuated from the danger area.
2. Necessary first aid procedures should be instituted, until emergency units can respond.
3. Notification to project management should be made activate and coordinate necessary rescue operations.
4. The emergency area should be secured to prevent unauthorized access.
5. Crew members should be designated to meet and direct emergency response units to the emergency site.
6. A crew roll call should be conducted to account for all workers in the area.
7. Damaged area should be inspected to ascertain the extent or damage and the need for rescue equipment.
8. Assist as directed with emergency response and rescue operations.
9. Do not give statements or discuss the occurrence without consulting with project management.
10. Assist the project emergency response team in their investigation.

Table 7–10 Bomb Threat Guidelines

1. Anyone receiving a bomb threat will immediately notify the project management. Because of the potential for a catastrophic occurrence any threat of a bomb will be treated as an emergency, until the validity of the threat is determined.
2. Employees and supervisors are not to touch, tamper or attempt to remove any suspicious or explosive device. Notify management at once when any device if found.
3. Evacuate all areas where a specific bomb threat exists. Evacuation must be accomplished in a orderly and controlled manner. After evacuation and assembly at a designated safe area, a crew roll call should be taken to assure that no one is unaccounted for.
4. Notify local authorities. It is their responsibility to conduct a bomb search and response.
5. Secure the perimeter of any bomb threat area to prevent unauthorized access.
6. Notify any affected business, residents or facilities who may be in danger.
7. Do not give a statement or discuss the occurrence without consulting project management.

Any employee involved in the making of a bomb threat will be immediately terminated and prosecuted to the full extent of the law.

Table 7–11 Severe Weather Guidelines

1. Projects are to prepare for severe weather problems, such as hurricanes, tornadoes, wind storms, etc., as soon as notice of a storm approach is received.
2. Each project should have an emergency phone contact list showing the home phone numbers of key project supervisors, personnel and subcontractors.
3. All loose trash should be removed from site and exposed areas swept clean.
4. Secure all loose materials or remove them from exposed areas. Flying materials during strong winds constitute a major hazard to personnel and property.
5. Banded materials like lumber should be secured to prevent movement. Chain castered equipment to columns or other secure locations.
6. Loose materials like gravel, nails, etc., should be place in tightly closed containers and secured or removed from exposed areas.
7. Exposed glass should be taped or covered with plywood.
8. Trailers should be secured with extra tie downs.
9. Order or reserve anticipated equipment or supplies needed for after storm clean up like pumps, generators, rope, wire, fuel, plastic, hose, electric cords, etc.
10. Lower and secure all crane booms to the ground. Secure all other equipment. Allow tower cranes to weather vane, follow all manufacture recommendations. Be sure to check tower crane support systems and the security of counterweights, wedges and clamps.
11. Fully fuel all equipment. Fill underground fuel tanks and safety cans. Have hand pumps available. Securely tie down all above ground tanks.
12. Fill water kegs with drinking water to insure an adequate supply for a few days.
13. Secure or remove barricades to prevent wind damage.
14. Take pre-storm pictures of the site and structures to assist in documenting storm damage. Take pictures of the area surrounding the job site, and your job site conditions after the job is secure. These pictures can assist in determining responsibility for storm damage to surrounding structures.

Table 7–11 (*Continued*)

15. Secure all ladders to the deck. Secure scaffolding to permanent structures, or dismantle. Secure rolling scaffolds to prevent movement.
    Secure scaffold decking and boards.
16. Halt shipment of nonessential materials to the job site when storm warnings are first received. This will prevent additional material from arriving after the job has been secured.
17. Make sure all site communications are in operating order. Fully charge all radios and issues radios as necessary to key personnel.
18. Remove all equipment from trenching and low lying areas and secure on high ground to prevent flood damage. This includes heavy equipment, welding machines, generators, compressors. etc.
19. Store expensive materials and equipment inside secure structures or protected areas.
20. Plan at what point of the storms intensity the electrical power should be shut down, and to what areas. Be sure that local utility repair crews are aware of any critical need for the job site.

    When using backup generators, be sure to plug equipment directly into a properly grounded generator. Do not connect the generator to the building circuits. Generators connected to building circuits will feed power directly into the power company utility grid.
21. Insure an adequate staff level to handle emergencies if the site or critical operations are to be maintained during storms. Adequate shelter, communications, provisions and emergency supplies will need to be on hand.
22. When a site is to be completely shut down consider arrangements for site security. Be sure that guard services in use will remain on site for the storms duration.
23. Consider arrangements with a local radio station to keep employees and subcontractors apprised of site closing, emergencies and return to work calls. Be sure that each employee receives instructions on what station to listen to and what steps to take in response to information broadcast.
24. Whenever possible, water flow should be diverted away from underground structures to prevent flooding.
25. After the storm has passed, inspect all areas of the project site to identify and abate any hazardous conditions before work commences.

EVACUATION PLANS

Notify all employees regarding the steps to be taken in an emergency evacuation of the workplace. The signals used for emergency evacuation, the designated evacuation routes, and the safe assembly point for each work crew must be detailed. Alternative safe assembly points for workers should be developed when the evacuation area must be widened.

Review emergency evacuation plans with each employee as part of a workplace safety orientation program. Follow-up training reinforcing emergency procedures should be incorporated into the site safety training materials.

Plan an emergency coordination center with proper communication equipment. Supervisors and employees assigned to assist in emergency procedures should know the location of the coordination center and the identities of the emergency coordination team members.

TRAINING

A sufficient number of employees must be trained to coordinate emergency procedures in the field. These designated employees and supervisors will assist in evacuation, roll-call information, and the operation of emergency equipment and facilities.

All employees will be trained in emergency evacuation plans, signals, safe assembly points, reporting procedures, and the emergency shutdown of equipment or operations. Employees should be trained in emergency procedures initially and at least annually thereafter.

First-aid, medical, fire-brigade, and rescue personnel when on site should perform practice drills periodically. Emergency plans should be periodically reviewed and adapted to the needs of the project and construction phase.

Community fire, police, rescue, and other emergency response groups should be given a project orientation and invited to participate in emergency procedure drills. Also, provide detailed site plans and copies of all emergency procedures to community emergency-response group coordinators.

Chapter 8

The Hazard Communication Standard

HAZARD COMMUNICATION STANDARD

This expanded manufacturing standard was extended to construction, service, retail, and other industries on May 23, 1988. This standard is at best a model of a good regulatory intent gone astray.

The Hazard Communication Standard in theory requires employers to educate workers on the problems involved with using dangerous chemicals in the workplace.

In reality, the standard attempts to apply a regulatory kind of one-size-fits-all doctrine. The same standard is applied to numerous kinds of industries without concern for the specific needs, conditions, or differences between industries.

Expansion of this Hazard Communication Standard is a glaring example of what happens when the courts interfere with the federal regulatory process, when agencies and personnel in the government bow to that interference, and when the people who draft the standard are uneducated about the industries they seek to regulate.

Unfortunately, this paperwork standard's main effect will not be to increase worker safety, but to generate numerous citations in the industries to which it is applied.

OSHA's recent change in the workplace inspection policy makes employer compliance with the Hazard Communication Standard the first item of business for all routine inspections.

The future for industries newly covered by this standard is evident in a recently released OSHA report. That report indicated that two and one-half years after the standard went into effect in the manufacturing industries most employers are still not yet in compliance. Missing paperwork accounts for over 60 percent of the citations issued.

REQUIREMENTS OF THE HAZARD COMMUNICATION STANDARD

This summary is provided to give a quick overview of the requirements of the Hazard Communication Standard. For more detailed information, please refer to the copy of the text of this standard, CFR 1926.59.

Employer Requirements

1. Conduct an inventory and list all hazardous chemicals used or stored in the work sites. Update that list as new chemicals are introduced. Maintain that inventory list on the work site. (See Figures 8–1 and 8–2.)
2. Collect Material Safety Data Sheets for each hazardous chemical on the inventory list. MSDSs must be maintained on the work site. (See Figures 8–3 and 8–4.)
3. Prepare a written Hazard Communication Program. That program must be maintained on the work site. (See Sample HCS Program.)

Prior to the start of work operations and as needed thereafter each contractor must complete these chemical information product cards and forward the cards to (Company Name). One card is needed for each chemical product used or stored on (Company Name) work sites. Whenever a new chemical product is brought on to the work site, a completed card and copy of the product MSDS must be given

Figure 8–1 Chemical Inventory Product Card

to the Site Project Manager. In addition, forward a copy of the completed inventory form and MSDSs to:

Company Name ______________________________

Company Address ______________________________

Contact Name ______________________________

MSDS # ______________ Product Name ______________

Supplier Information: Phone No. (__) ______________

Name ______________

Address ______________

______________ Label Warnings ______________

____

____

Hazardous Ingredients Listed On Label or MSDS:

1. ______________ 6. ______________

2. ______________ 7. ______________

3. ______________ 8. ______________

4. ______________ 9. ______________

5. ______________ 10. ______________

Quantity ______________ Location ______________

____

Classification:

Hazardous ______________ Nonhazardous ______________

Consumer Product ______________

Date ______________ Signature ______________

Figure 8–1 ***(Continued)***

CHEMICAL INVENTORY SHEET **Page#____**

Product Name	Manufacturer Information	Chemical Ingredients	Amount	Location	Class

Figure 8–2 Chemical Inventory List Sheet

Date:

To: Chemical Supplier or Distributer

Street Address

City, State, Zip Code

Dear :

In order to comply with OSHA's Hazard Communication Standard, we are requesting Material Safety Data Sheets for the following chemicals/substances listed below:

1.

2.

3.

Your prompt attention to this matter would be appreciated.

Sincerely,

Project Manager

cc: file

Figure 8–3 MSDS Request Form

Date:

Administrator

Health Standards Division

Occupational Safety and Health Administration

200 Constitution Avenue

Washington, D.C. 20210

Dear :

As required by the Hazard Communication Standard (company name) has requested Material Safety Data Sheets from ____________, see attached request letter(s).

As of this date, the supplier has not forwarded the material requested or responded to our requests. Please investigate this matter and provide us with assistance in obtaining the necessary MSDS(s).

Sincerely,

Project Manager

cc: supplier
file

Figure 8–4 Notice to OSHA of nonreceipt of MSDS

4. Ensure that containers for all chemicals are properly labeled.
5. Exchange hazardous chemical information with other employers on site whenever there is a potential for worker exposure to hazardous chemicals. (See Figure 8–5, Subcontractor Notification Letter.)

Date:

To: All Subcontractors

From: Company Name

Re: Hazard Communication Standard

The Occupational Safety and Health Administration requires all employers in the construction industry to comply with the Hazard Communications Standard, 29 CFR 1926.59.

The purpose of this standard is to provide workers with information on the hazards of chemicals they may work with, or be exposed to, on the job site. (company name) written program, chemical list, and MSDS file is maintained on project sites and at our main office.

All subcontractors, as a condition of employment, must provide copies of the following to (company name) for all projects and work sites.

1. Your Company's Written Hazard Communication Program
2. Your Company's Chemical Inventory List
3. Copies of Material Safety Data Sheets for all hazardous chemicals or substances you use or store on company work sites.

Figure 8–5 HCS Subcontractor Notification Letter

4. The name of your company's hazard communication coordinator and maintenance location of your company's HCS program, chemical list, and MSDS file for all company projects.

Under the Hazard Communication Standard, *each employer* is solely responsible for:

- Providing required training to all employees,
- Compliance with the standards requirements,
- The exchange of information on hazardous substances used or stored in the work sites with other employers, and
- Employees access to chemical information.

(company name) assumes no responsibility or liability for any other employer's compliance with this standard's requirements.

Figure 8–5 (*Continued*)

6. Each employer must provide information to employees on the requirements of the Hazard Communication Standard and the hazards of materials they may be exposed to. (See Figure 8–6, Employee Notification Letter, and Figure 8–7, Supervisor Notification Letter.)

Date:

To: All Company Employees

From: HCS Coordinator

Re: The Hazard Communication Standard

(company name) is committed to providing our employees with a safe and healthy work environment. In compliance with the federal Hazard Communication Standard, (company name) has instituted a company hazard communication program.

Figure 8–6 HCS Employee Notification Letter

The purpose of this hazard communication program is to provide information to workers on the hazards of chemicals they use or may be exposed to in the work place.

(company name) HCS Program consists of five primary elements:

1. A written company hazard communication program.
2. A chemical inventory list of chemicals used or stored in the work place.
3. A file of material safety data sheets for those materials.
4. A training program to provide information to workers.
5. The exchange of chemical information with other employers on (company name) jobs.

Employees have the right to be informed about the chemicals they work with and to review (company name) HCS program, chemical inventory sheets, and material safety data sheets.

Employees also have an obligation to be alert to the hazards of chemicals on the job, review MSDSs for chemicals in use, follow safe work practices, and use appropriate personal protective equipment.

All subcontractors of (company name) will be required to adhere to the provisions of the federal Hazard Communication Standard, and (company name) HCS Program. For more information, please contact your job site supervisor.

Figure 8–6 ***(Continued)***

7. Provide employees with the opportunity to review the HCS written program, chemical lists, and material-safety data sheets.(See Figure 8–8, Employee Information Request Form.)

What Is a Hazardous Chemical Under OSHA?

1. A health hazard is considered to be any chemical, substance, or mixture that may adversely effect the health of

Date:

To: All Company Supervisors

From: HCS Coordinator

Re: (Company Name) HCS Program

In accordance with the federal Hazards Communication Standard, (company name) has instituted a Company Hazard Communications Program.

That program consists of a written HCS policy, chemical inventory sheets, a file of material safety data sheets, and an employee training program.

Our employees have the right to review our company's HCS policy, chemical inventory sheets, and the material safety data sheet for any material they work with.

Each supervisor will be an integral part of our hazard communications effort. Supervisors will be responsible for the implementation and coordination of HCS efforts in the areas under their authority.

Crew supervisors will utilize HCS related toolbox talks in their crew safety meetings. Crew supervisors will insure that their work operations are reviewed for chemical hazards and the crew briefed on any hazards prior to the use of a particular material. Project Managers will maintain the written program, chemical inventory,

Figure 8–7 HCS Supervisor Notification Letter

and MSDS file on the job site and update the inventory list as necessary.

The Project Manager will also interface with subcontractor supervisors to exchange needed information. All subcontractors of (company name) will be notified of their obligations under the Hazard Communication Standard.

The task of complying with this standard is not an easy one, but with your help it can be done. Join with me as we work towards the reduction of chemical related injuries and illnesses on all company projects.

Any HCS questions, suggestions, or concerns should be directed to the Project Manager.

Figure 8–7 ***(Continued)***

a person with any short- or long-term exposures. Hazards can include any chemicals that are:

a. carcinogens
b. toxins
c. mutagens
d. teratogens
e. irritants
f. sensitizers
g. corrosives

2. Further, any chemical that presents a physical hazard to any person who may be exposed is considered a hazard. These include chemicals that are:

   a. flammable
   b. explosive
   c. combustible
   d. oxidizers
   e. reactive

3. A hazard is also considered to be any by-product of a work operation that has any of the hazards listed above. This includes exposures to common construction materials like wood dusts and welding fumes.

EMPLOYEE INFORMATION REQUEST FORM

As an Employee of (company name), I am requesting a copy of the Material Safety Data Sheet(s) for the following chemicals, substances, or materials used or stored on (*name of job site*):

1. ____________________

2. ____________________

3. ____________________

____________________ ____________________

Date Employee's Signature

I acknowledge that on ________ I received or reviewed material safety data sheets for the items requested above.

____

Employee's Signature

Figure 8–8 Employee Information Request Form

4. Any consumer product with warnings on the label or MSDS, not used in the same manor or quantity as a normal consumer would use them, is also considered hazardous.

Hazard Determination

Under this standard, employers have no obligation to conduct a hazard determination analysis of any chemicals they use or store. Employers may rely on the hazard determination made by the chemical manufacturer or supplier.

A chemical product, substance, or mixture is considered hazardous when the manufacturer places any warning language on the product label or MSDS.

Manufacturers, distributors, and suppliers of chemical products must supply material safety data sheets to purchasers of their products.

Effective Dates

Each construction employer must comply with Hazard Communication Standard as of March 1989. Employers who under federal law are required to maintain material safety data sheets must also comply with reporting requirements under the Community Right To Know Act of 1986.

This sample written HCS program is provided to assist employers in complying with the paperwork requirements of the standard. Each employer should review and modify this program to fit the company's specific needs.

SAMPLE HAZARD COMMUNICATION WRITTEN PROGRAM

The OSHA Hazard Communication Standard 29 CFR 1926.59 requires a company to inform its employees of the hazards associated with use of and exposure to chemicals used in the workplace.

This program has been prepared to comply with all requirements of the Federal OSHA standard 1926.59 and to ensure that information necessary for the safe use, handling, and storage of hazardous chemicals is available to all workers.

This program includes guidelines on the identification of chemical hazards, and the preparation and proper use of containers, labels, placards, and other types of warning devices.

Chemical Inventory

The company maintains an inventory of all chemicals it uses in its work place. A chemical inventory list and written hazard communications program are available from the company or project office or by contacting the hazard communication program coordinator.

All chemicals brought onto the work sites must be included on the project chemical inventory list. To ensure compliance with federal requirements, any chemical brought onto the job site should be added onto the Chemical Inventory Log Sheets maintained at the Project Superintendent's office.

Labeling

All chemicals on site must be stored in either their original or approved containers, with a proper label attached. Any container not properly labeled should be given to a *supervisor* for labeling or disposal.

Workers may dispense chemicals from original containers only in small quantities intended for their immediate use. Any chemical left after work is completed must be *returned to a proper storage area* for handling. No unmarked containers of any size are to be left in the work area unattended.

Material Safety Data Sheets

Employees may request a copy of the material safety data sheet (MSDS) for any chemical or substance considered hazardous with which they work. Requests for MSDSs should be made to *the employer or the crew supervisor* using the proper form. When such a request is made, standard chemical references may also be used to provide chemical safety information.

Employees have the right to review a copy of their employer's written Hazard Communication program and chemical inventory sheets. All such requests should be made in writing to your supervisor or the Project Superintendent.

Employee Training

All employees will be trained by their employer to work safely with hazardous chemicals. Employee training will include:

1. Methods that may be used to detect a release of a hazardous chemical(s) in the workplace.
2. Physical and health hazards associated with chemicals.
3. Protective measures that should be taken if exposure occurs.
4. Specific safe work practices, emergency release, or spill responses and the use of personnel protective equipment.

5. Details of the Hazard Communication Standard, including:
   —label and warning systems;
   —explanation of Material Safety Data Sheets;
   —how employee can obtain the necessary hazard information on the job site.

Employees will be trained initially, prior to working with hazardous chemical and as necessary to provide additional chemical information. All employees will receive a package of written material on the hazard communication program. Employees must return the form found in the information package indicating that they have received and are familiar with the HCS program contents.

Personal Protective Equipment (PPE)

PPE is available from *your supervisor*. Required personal protective equipment must be used at all times as needed. Failure to use required PPE will result in employees being subject to disciplinary actions up to and including discharge. Using the Personal Protective Equipment provided by your company is for your benefit and protection.

Emergency Response

Any incident of overexposure or spill of a chemical substance must be reported to the superintendent at once.

Area superintendents will be responsible for ensuring that proper emergency response actions are taken in all leak/spill situations.

Hazards of Nonroutine Tasks

Supervisors must inform employees in their workplaces of any special tasks that may arise that would involve possible exposure to hazardous chemicals.

Review of safe work procedures and the use of required PPE should be conducted prior to the start of such tasks. Where necessary, work areas will be posted to indicate the nature of the hazard involved, by the crew using hazardous chemicals.

No work may be performed in any confined or poorly ventilated area without proper ventilation, protective equipment, safe work practices, and supervision.

Informing Outside Contractors and Subcontractors

All outside contractors and subcontractors must adhere to all provisions of the company's hazard communication program and the Hazard Communication Standard.

Information regarding hazardous chemicals known to be present in the workplace must be exchanged between employers prior to the start of such work. Each employer will be responsible for providing necessary information to their employees.

All outside contractors must obtain a copy of the company hazard communication program and provide the company with a copy of their company's written HCS program, chemical lists, and MSDSs for all hazardous chemicals brought into the workplace.

CONDUCTING A CHEMICAL INVENTORY

The Federal Hazard Communication Standard requires all employers to maintain on site a list of hazardous chemicals known to be present in the workplace. That list must identity each chemical or substance by one of the names or identifiers used on the manufacturer's material safety data sheet.

Chemical inventory lists may be compiled for the entire workplace or for each workplace where hazardous chemicals are present, used, or stored.

Under the current OSHA HCS standard only specific chemicals are exempt from coverage. These include chemicals or substances that are:

1. Labeled as pesticides.
2. Foods, food additives, food, or cosmetic and coloring agents and prescription or patent drugs, regulated under the Food and Drug Administration.
3. Distilled spirits, wine, or malt beverages for noncommercial use.
4. Hazardous waste materials regulated by the Environmental Protection Agency.

5. Tobacco or tobacco products.
6. Solid woods or wood products, but not treated woods, wood dusts, or chemicals used to treat or coat wood.
7. Consumer products regulated under the Consumer Product Safety Commission used in the same manner and creating the same exposure in the workplace as they would for normal consumer use.
8. Articles, any solid material that will not be altered in the workplace to produce fumes, dusts, mists, vapors, or gases. Also, a solid material that does not release more than a few molecules of any hazardous substance.

Most exempted chemicals will have minimal impact on construction operations.

HAZARDOUS CHEMICALS—OSHA

Any chemical labeled by a manufacturer or any recognized authority as hazardous, unless exempted, is covered under the scope of the Hazard Communication Standard.

Under the Hazard Communication Standard, there are no exposure limits set. Therefore, any quantity of any hazardous chemical (more than a few molecules) is sufficient to trigger all HCS requirements.

OSHA defines hazardous chemicals as:

1. Any chemical listed as toxic.
2. Any chemical listed as a carcinogen or a potential carcinogen.
3. Any chemical labeled as corrosive.
4. Any chemical irritants or sensitizers.
5. Any by-product produced with any of the effects listed above.
6. Any consumer product used in a manner or concentration outside of normal consumer uses.

Consumer Products

OSHA's coverage of consumer products complicates the application of the Hazard Communication Standard to the workplace. Contractors are advised to treat consumer products as potentially hazardous materials by listing them on the chemical inventory and obtaining a MSDS for each product.

By-Products

OSHA includes by-products with any hazardous chemical effects under the scope of the Hazard Communication Standard. Employers will need to be aware that many common substances could be considered hazardous, including wood dusts, welding fumes, concrete dust, and airborne mineral wool fibers.

Employers should review work operations to determine potential by-product production from chemical, mechanical, and thermal actions.

CHEMICAL INVENTORIES—GETTING STARTED

A wall-to-wall inspection of the work sites and storage areas should be conducted to develop a listing of all products stored or in use.

The following information copied from the container labeling should be placed on the chemical inventory product cards found in Figure 8–1.

1. Product name.
2. Chemical name(s).
3. Manufacturer's name and address.
4. Container size.
5. Number of containers (count).
6. Location.
7. Product I.D. numbers.
8. Any product warnings on the label.

Any materials that cannot be identified or have improper, damaged, or missing labels should be removed to a secured area for proper disposal in accordance with federal, state, and local law.

Once all the chemical product inventory cards are completed, they can be utilized to compile a chemical inventory list. A blank Chemical Inventory Sheet form is provided in Figure 8–2.

A manufacturers' mailing list can be compiled from the inventory list to assist in requesting material safety data sheets for all chemicals, products, or substances used on the job. See sample MSDS request letter in Figure 8–3 and nonreceipt of MSDS notice in Figure 8–4.

OSHA has indicated that generic MSDSs are acceptable in place of product specific MSDSs. OSHA, however, will hold the employer responsible if generic MSDSs are incomplete.

Once the chemical inventory sheet is compiled, employers must ensure that new or additional MSDSs are ordered as needed and chemicals, products, or substances are added as they are brought into the workplace.

A master chemical inventory for a company can be compiled and utilized as the chemical inventory lists for a particular project even though the master inventory will be more extensive than the normal project inventory list.

Chemical inventory lists must be maintained at each work site and made available upon request during OSHA inspections. Some states require that inventory lists and MSDSs be posted at the job site. Refer to your State Right to Know Law for additional information.

UNDERSTANDING MATERIAL SAFETY DATA SHEETS

Introduction

Material safety data sheets, or MSDSs, are developed by chemical manufacturers, distributors, or importers to provide chemical information about their products.

There is no standardized format or language for MSDSs, which makes using them somewhat difficult. But they all provide the same type of information grouped in the same general categories.

For purposes of our discussion, we will discuss the typical MSDS format used by the Occupational Safety and Health Administration.

Section I—Identity

This section tells you about the terms associated with the material and information about the supplier or manufacturer. This identity and contact information can be used to obtain additional information from the supplier or others about a product.

Common Terms

Chemical name	What this product is called by chemists.
Trade name	What the manufacturer or supplier calls this material.
Generic name	What the common name is for this product.
Other synonyms	What other names this material is known by.
Supplier ID	Who makes this product and what their address is.
Emergency phone number	How to obtain additional information about this product.

Section II—Hazardous Components

The hazardous chemicals in a product are listed in this section. Additional names for the chemical(s) are also listed.

Established exposure limits for the material or its chemical components are listed along with the percentage of the individual components.

The exposure limits are important because they tell you what is considered a safe amount of that chemical to which you may be exposed. When exposures exceed that safe exposure level, you may be at risk.

Common Terms

OSHA PEL	The permissible exposure established by the Occupational Safety and Health Administration.
ACGIH TLV	A nonmandatory exposure limit established by the American Conference of Governmental Industrial Hygienists.
NIOSH REL	The National Institute of Occupational Safety and Health recommended exposure limit.

C	The ceiling or maximum exposure limit that should not be exceeded.
STEL	Short Term Exposure Limit, a limit that is excepted for exposures that do not exceed 15 minutes (also called the excursion limit).
ppm	The parts per million.
mg/m^3	Milligrams per cubic meter.

Section III—Physical and Chemical Characteristics

The appearance of a material and its characteristics are described in this section. These characteristics can help you identify an unknown material or chemical.

Common Terms

Boiling Point (BP)	The temperature at which a liquid boils and turns into a gas.
Vapor Pressure (VP)	The force exerted by a liquid as it changes into a gas. Usually measured as millimeters (mm) of mercury (Hg). The higher the vapor pressure, the more the material will evaporate and enter the air.
Vapor Density (VD)	Weight/volume of a gas as compared to the weight/volume of air. Air is considered as 1. If the vapor density is less than 1, the gas will rise. If the vapor pressure is more than 1, the gas will sink.
Evaporation Rate (ER)	The speed at which a liquid converts to a gas when compared to butylacetate (BUAC). Butylacetate equals

	1. A high ER means the material may present an inhalation hazard.
Specific Gravity	The weight per volume of material compared to water, where water equals 1. Materials with a specific gravity above 1 sink, and those below 1 float in water.
Melting Point (MP)	The temperature at which the material turns from a solid to a liquid.
Solubility	The amount of material that will dissolve in water.
Appearance and Odor	How this material looks and what it smells like. Odor is not a good method to identify an unknown chemical. Odor is unreliable and the material may be harmful if inhaled.

Section IV—Fire and Explosion Data

This section will tell you about the fire and explosion hazards associated with a material. Precautions for fighting a fire and fire-related chemical and health hazards are also listed.

Common Terms

Flash Point	The temperature at which this material will give off vapors that might ignite. If the material has a flash point less than 100 degrees Fahrenheit, then it's called flammable. If the material has a flash point greater than 100 degrees Fahrenheit, it's called combustible.
Lower Explosive Limits (LEL)	The minimum concentration of this material in air that will explode.

Upper Explosive Limits (UEL)	The maximum concentration of this material in air that will explode.
Flammable Limits	The percentages minimum to maximum of this material in air that may burn or explode.
Extinguishing Media	What can be used to put this fire out.
Special Precautions	What procedures or equipment should be used in a fire situation.
Unusual Hazards	Whether or not this material will give off dangerous gases or create dangerous materials in fire and explosion situations.

Section V—Reactivity Data

This section will tell you how a chemical may react when mixed with other chemicals. Reactivity data outlines potentially dangerous situations.

Common Terms

Stable	This material resists chemical changes.
Unstable	This material can be dangerous because it easily undergoes chemical changes.
Incompatibility	Lists what type of chemicals will create dangerous reactions with this material. This chemical should never be used, stored, or mixed with the incompatible chemicals listed.

Hazardous Decomposition	What toxic or dangerous products can be produced by this material. Often these decomposition products can be more dangerous than the original material.
Hazardous Polymerization	Whether or not this product can combine its molecules into long chemical chains. The formation of long-chain molecules is associated with runaway explosive reactions.
Conditions to Avoid	Usually refers to heat but may list any conditions that contribute to the material's instability.

Section VI—Health Hazard Data

This section identifies a chemical in this material that can cause injury. Important first aid procedures are listed in this section.

Common Terms

Routes of Entry	How the material can get into your body: *Inhalation* means breathing the material in; *ingestion* means swallowing the material or substances contaminated with this material; and *absorption* means the material passes through the skin or tissues.
Skin	Notation to indicate that this material can be absorbed through the skin.

Acute	Short-term health effect.
Chronic	Long-term health effects.
Carcinogenicity	The linking of this material to the formation of cancer. Materials may be listed as a known or suspected cancer-causing agent in animal or human studies.
NTP	The National Toxicology Program, aprogram to identify carcinogens.
IARC	The International Agency for Research on Cancer. They publish *The IARC Monographs,* a list of carcinogens.
OSHA Regulated Carcinogen	A carcinogen listed in an OSHA regulation or on the 1910.1000 list.
Signs and Symptoms	The most common or most likely effects of exposure to this material. Because individual reactions to exposure vary, not all effects can be detailed.
Medical Conditions Affected	What preexisting conditions can be aggravated or worsened by exposure to this material.

Section VII—Precautions for Safe Handling and Use

This section will tell you how to use and store this material safely. The cleanup of spills and safe and legal disposal methods are detailed.

Emergency procedures for spill and leak situations are discussed along with any special precautions you should take. On some MSDSs, this section will be broken down into storage and spill and leak sections.

Section VIII—Control Measures

This section will discuss the use of personal protective equipment associated with this material. Recommended safety equipment and clothing are listed here. Safe work practices and hygiene methods outline the steps to be used when working with this material and cleanup procedures for workers.

The type of engineering controls recommended for controlling exposures to this product are discussed. This section contains vital information on protecting yourself from chemical injuries.

Common Terms

Approved Respirator	Using a respirator designed for the type of chemical exposure involved.
Filter Mask	A type of respirator using a filter media to control exposure. Usually used for dust or fiber exposures.
Canister Respirator	A type of respirator that uses a chemical absorbent to control a specific type of chemical exposure.
Forced Air Respirator	A respirator that supplies an uncontaminated source of air to breath.
SCBA	Self-Contained Breathing Apparatus.
Respirator Fit Testing Program	A procedure required by OSHA regulation to ensure that the worker is suited for respirator work and using a properly fitted and maintained respirator.

Natural Ventilation	Dilution of the chemical concentration by normal air movements.
Mechanical Ventilation	Forced ventilation via fans, blowers, or other means.
Isolation or Enclosure	Control of chemical exposure by specific engineering means.

Additional Information

Although material safety data sheets may differ in their format, they will all contain the type of information outlined. When you have additional questions or need more specific information on the MSDS or materials you work with, contact your supervisor.

EMPLOYEE TRAINING

Each employer under the Hazard Communication Standard is required to provide training to the company's employees. That training includes the recognition and avoidance of the hazards of chemicals they may be exposed to and the provisions of the Hazard Communication Standard.

Employers must train employees before their initial work assignments and as new chemical exposures occur in the work site.

All employees should receive a hazard communication information package and HCS orientation training prior to their initial work assignments.

Additional training should be provided at the field level to reinforce that initial training and to deal with specific hazards when necessary. This training can be conducted in conjunction with a company's normal safety-training programs.

Employee Training Documentation

Each employee, as part of initial training, should be required to fill out and return a receipt form to document initial HCS training and information. A sample employee HCS training receipt form is found in Figure 8–9.

EMPLOYEE RECEIPT FORM

I have carefully and thoroughly read (company name) hazard communication written program. I have received or reviewed a copy of that program and understand its requirements. I agree without reservation to follow the company's HCS program.

I also understand that I have the right to review the company's chemical inventory lists and the material safety data sheets for any chemical I may use or be exposed to. I have been informed of the person to contact and the location where this information is maintained for my job site.

I agree to follow safe work procedures when dealing with chemicals and to use proper personal protection equipment when necessary.

____________________	____________________
Date	Employee's Signature

Figure 8–9 Employee Receipt of HCS Training Form

Each time additional safety training is conducted, supervisors or the trainer will forward a training attendance list to the company. That list will contain the printed name of each attender, date of training, and the subject(s) covered.

Training Documentation Records should be maintained on the project's main site for the duration of the project work schedule.

Documentation of an employee's training assists the company in defending against an OSHA citation alleging inadequate employee training. Under the language of the Hazard Communication Standard, OSHA compliance officers will be able to judge the adequacy of an employer's HCS training. Compliance officers will make very subjective determination by questioning workers about their knowledge of the standard and the hazards of chemicals. Employers with adequate documentation of all HCS training will be able to successfully prove their "good faith effort" to comply with this regulation.

Record Keeping Requirements

At the end of the project, training records, MSDSs, and any exposure monitoring records should be forwarded to the company's main office for storage.

Material Safety Data Sheets are considered part of the employee's medical records and must be maintained for a period of 30 years.

SARA COMMUNITY RIGHT TO KNOW REGULATIONS

The Superfund Authorization and Redevelopment Act (SARA) of 1986 contains the Community Right To Know Act.

This Federal Emergency Planning and Community Right To Know Act, sections 311 and 312, place reporting requirements on anyone who, under federal law, must maintain material safety data sheets.

Contractors now covered by the Hazard Communication Standard are required to submit appropriate chemical hazard information to the local emergency planning committee, state emergency response committee, and the local fire department.

The Federal Environmental Protection Agency (EPA) has jurisdiction over SARA and has established Threshold Planning Quantities (TPQs) for hazardous chemicals.

If you use, maintain, or store hazardous chemicals in quantities above the established TPQ, you must submit copies of MSDSs or a chemical inventory list. You must also file annual reports of hazardous chemical maximum quantities and average daily quantities.

Sara Section 311—Reporting Requirements

If you maintain one or more hazardous chemicals in quantities exceeding the EPA limit, you are required to submit MSDSs or a chemical inventory list.

MSDSs must be sent to state and local emergency planning groups and the local fire authority for each hazardous chemical above EPA limits.

If you choose to submit a chemical inventory list, it must be grouped by categories of health and physical hazards. The chemical name or common name of the product and any hazardous ingredients must be provided.

A sample SARA reporting letter is contained in Figure 8–10.

Sara Section 312—Annual EPA Reporting

When you maintain more than the minimum EPA quantity of a hazardous substance, you are required to file an EPA Emergency and Hazardous Chemical Inventory Form. Also called the Tier I Report, this form includes an estimate of the maximum amount of hazardous chemicals present in the workplace at any time. Location information for chemical use or storage and estimates of average daily amounts present are required.

This report is required for each project, and storage or use location. Contractors exceeding the TPQ limits on any project or job site would be required to submit a report form for that location.

Sara Compliance Assistance

To assist employers in compliance with the SARA requirements, the Environmental Protection Agency provides the following:

1. EPA Chemical Emergency Preparedness Program Hotline (CEPP) 1-800-535-0202. The CEPP operates from 8:30 A.M. to 4:30 P.M. Eastern Standard Time. The Washington, D.C., CEPP number is 202-479-2449.
2. An EPA list of Threshold Planning Quantities for hazardous chemicals and Tier I and Tier II report forms and instructions are available through the CEPP or the local EPA office.

Additional compliance assistance for SARA is available from the following:

1. The Superintendent of Documents, U.S. Government Printing Office. Copies of SARA and the Community

Date:

To: State Emergency Response Planning Agency,
Local Emergency Response Group,
Local Fire Authority

Dear :

The Community Right To Know Law of 1986 (SARA sec. 311 & 312) requires that any employer who must maintain Material Safety Data Sheets under a federal law must provide an annual notification to the proper authority.

Our company, under the Hazard Communication Standard maintains MSDSs. Therefore, we are submitting our company chemical inventory list for the following job sites. Copies of the MSDSs for these materials are available upon request.

Sincerely,

Project Manager

cc: file

Figure 8–10 Sara 311 Notification Letter

Right To Know Act are available for a nominal fee. Additional publications are listed in a GPO publications catalog, available upon request.

2. The Small Business Ombudsman Hotline (SBOH) provides compliance assistance and information. This non-governmental service is in operation 7:30 A.M. to 4:30 P.M. workdays, Eastern Standard Time. The SBOH can be reached at 1-800-368-5888.
3. Each state under the Community Right To Know Act of 1986 was required to create a state emergency response commission within the local emergency planning districts. Your State Emergency Response Commission can provide information about your state program and law.

Chapter 9

Accident Investigation

ACCIDENT RECORD KEEPING

The Occupational Safety and Health Act requires all employers to maintain a log of recordable occupational injuries and illnesses. Therefore, any employee who receives medical treatment, suffers a loss of consciousness, a restriction of work motion, time lost from work, or an injury-related transfer to another job must have the accident details recorded on the company's OSHA log (called the OSHA 200 Log).

The OSHA 200 Log must be maintained on each job site and made available for review during an OSHA inspection. During February of each year, the OSHA 200 Log must be posted at the work site.

Accident Investigation

Any accident involving a serious injury to any company employee or the employee of a subcontractor or trade contractor to the company must be investigated. All accidents involving property damage, whatever the cause, must be investigated.

The purpose of accident investigation is fact finding, not fault finding. Accident investigation ensures that the causes of an accident are known and that the hazards or conditions that caused the accident can be eliminated and additional injuries prevented.

Preliminary accident investigation is initially conducted by the crew supervisor. The crew supervisor first ensures that emergency first-aid treatment is available for all injuries. Notification can then be made to project management. To prevent further injuries, and assist with the investigation, secure the accident site until the investigation is completed.

It is the supervisor's responsibility to ensure that any physical evidence related to the accident is preserved to the fullest extent possible. The preliminary investigation determines the basic circumstances surrounding the accident and the identity of any witnesses. Photographs of the accident scene are a valuable investigative tool.

As a part of the company's safety policy, all supervisors and employees must cooperate fully in any company investigation. When outside investigative services are utilized, their on-site actions must be coordinated through a company representative.

Remember, the real purpose of an accident investigation is to establish all the relevant facts related to the accident occurrence, so that appropriate corrective actions can be taken. Property damage and near-miss accidents are also important. Careful investigation of these occurrences can pinpoint potential problems that could lead to injuries.

The difference between a near-miss accident and a serious injury is really just a matter of luck.

Conducting an Accident Investigation

The following considerations will assist you in conducting an investigation of any injury, property damage, or near-miss occurrence:

1. *The Five W's.* Who? What? When? Where? and Why? Keep these words in mind when conducting an investigation. Answering all these key questions will ensure that your report is thorough.
2. *Position.* There are always two sides to any question. Therefore, the physical and the emotional aspects of an accident must be considered by the investigator. Always ask questions like where was the injured worker, where were the witnesses, what are their opinions, and what are the facts? A good investigator listens to both sides of the story before deciding.

3. *Initial report.* Sometimes, you cannot find all the facts before you file your initial report. However, don't delay; file your report as soon as possible. Your initial report sets the company's accident procedures in motion. A follow-up report can be filed later.
4. *Evidence.* Evidence comes in all sizes and shapes, so don't overlook anything! Evidence must be preserved in order to be useful.
5. *Supplemental material.* Use attachments to make reports more understandable and complete. Attachments to your report could include the use of:
   A. Diagrams: A simple drawing showing the relationship of all the pertinent elements at the accident scene, noting measurements, distances, sizes, etc.
   B. Photographs: Every photograph should be accompanied by a brief description of what is shown, when it was taken, who took it, and why. Additional notations should include information on the camera, film, and lens settings.
   C. Statements: Any written or recorded statements of witnesses taken should accompany the report. A witness statement can be as simple as having the person making the statement write it, date it, and sign it.
   D. Physical objects: Any defective materials, tools, etc., related to the accident should be tagged and secured. Attach an inventory of all such physical objects to the report.
   E. Additional information: Reports made by police, fire department, paramedics, etc., are valuable information for a thorough investigation. Whenever possible, obtain copies of these reports. If these reports are not available, note the names and identity of those making such reports.
   F. Miscellaneous: Attach any additional forms or backup material that may be useful to the investigation, including maps, blueprints, equipment documentation, etc. Remember, when an accident is reviewed after the fact, only the information included in your report is available. Too much information is better than too little.

6. *Third parties and subcontractors.* The most common and costly error in construction accident reporting is not identifying third parties involved in the accident. These include subcontractor employees, vendor employees, or members of the public. The involvement of any person or entity in the accident should be identified on the report.

As a part of the company's safety policy, subcontractors must notify the project manager immediately following any accident involving injury or property damage. A copy of the first report of the injury and all accident investigation records should be forwarded to the project manager's office.

Third-party suits are difficult for a construction company to defend against and are potentially costly. Documentation of the accident will assist in protecting against these liabilities.

FIRST AID

To provide protection for your company and your employees, make arrangements with a local infirmary, clinic, or hospital to provide emergency care for all injured employees.

When available, utilize local emergency response facilities such as fire, police, ambulance, and rescue as part of your site first-aid plan.

Post the telephone numbers for the hospital and local emergency-response facilities on all job sites.

OHSA requires, at a minimum, that one or more employees on each shift be trained and hold a current certification card in first aid. Providing on-site emergency first-aid facilities helps the employer in controlling accident costs.

OSHA requires that first-aid supplies and emergency equipment be available on each project site. All supplies are to be of an approved type and be readily available to the work in progress. A First-Aid Checklist is provided in Table 9–1.

Supervisors are responsible to ensure that all employees under their control receive treatment for any injury.

DEFINING ACCIDENTS

What is an accident? Experts define an accident as, "an undesired event that results in physical harm to people or damage to property."

Table 9–1 First-Aid Checklist

1. Are the emergency phone numbers clearly posted on site?
2. Are adequate, approved first-aid and emergency treatment supplies readily available?
3. Are on site first-aid facilities or one or more employees certified in first aid available on each shift?
4. Are medical personnel readily available for consultation and advice on employee health matters?
5. Have arrangements been made for local emergency response and treatment in case of injury?
6. Where employees are potentially exposed to harmful corrosive materials are quick drenching and flushing facilities provided?

NOTE: Some states and jurisdictions have laws concerning medical practices which limit the type of first-aid treatment administered by laypersons. Be sure to check your local laws.

We might broaden that outlook to include a definition of incidents as, "an unwanted event that could or does downgrade the efficiency of a business operation."

Only a small percentage of incidents result in accidents, but all incidents have a negative effect on business. The difference between an accident and an incident is opportunity. The worker being under the falling load is equal to an accident. The worker being three feet away from the falling load is equal to an incident. Incidents are usually termed near misses. Near misses can be just as important as accidents because they both signal malfunction at the job site.

No matter how conscientious an investigator is or how extensive your company's accident investigation program is, you cannot begin to investigate until you are made aware of the occurrence.

Why then are some accidents never reported?

1. There is no established policy.
2. To protect safety records.
3. To avoid production-time loss.
4. To avoid blame; self-protection.
5. To avoid costs or red tape.

Companies can assure good reporting by taking a few suggested steps:

1. Establishing reports and procedures in writing.
2. Training all employees in the need for accident reporting and *all* reporting procedures.
3. Training accident investigators to ensure accurate, complete, and unbiased reports.
4. Taking action on all accident reports; actions speak louder than words.

ACCIDENT INVESTIGATION TECHNIQUES

The first reason for accident investigation is to uncover the underlying cause of the accident. Was it a result of an unsafe work practice that could have been prevented? Was the problem the violation of regulations? Did physical and environmental factors contribute to the accident? Was there a tool or equipment problem?

All accidents involve one of three primary factors: human, situational, or environmental.

Human factors relate to what a person does or fails to do that contributes to an accident. These unsafe acts depart from normal, safe work procedures or violate safety regulations.

Situational factors involve physical or site hazards that contribute to an accident cause including inadequate planning, defective designs, substandard work, and missing or required safety materials.

Environmental factors fall into three broad categories: physical, chemical, or biological factors.

The second reason for accident investigation is to identify the actions that can be taken to prevent similar occurrences.

Once identified, hazardous conditions, unsafe acts, or environmental concerns can be addressed and abated. Without understanding the necessary steps to decrease accidents, accidents will continue to occur.

The third reason for accident investigation is to establish the facts involved in the occurrence. Compensation and litigation are unfortunately facts of life. Unless a company is prepared, it will not be able to protect itself or its workers.

Accident investigation provides insight into problems that indirectly contribute to the occurrence of the accident. This information is vital in accident-reduction efforts.

A detailed investigation is likely to determine underlying causes involved. For example, although a worker may have slipped on a walkway, thorough investigation might reveal other causative factors, including poor housekeeping, improper maintenance, an inadequate supervisor, faulty equipment, etc.

Investigation of accidents provides information related to the direct and indirect costs involved.

Accident costs can be divided into two categories—direct and indirect. Direct costs are those covered under insurance (medical bills, premium adjustments, property and liability costs). Indirect costs are those uninsured losses, such as replacement costs, lost production time, administrative costs, court costs, and court awards.

Indirect costs exceed direct costs by a ratio of from 2 to 1 through 17 to 1. Therefore, for every dollar in direct accident costs, the total costs of accidents can be multiplied 2 to 17 times.

Despite a popular belief, the purpose of accident investigation is fact finding, not fault finding. There is a common tendency to place the full blame of an accident upon the injured person. However, if human error is not the sole factor in the accident those hazards and causative factors involved will go unobserved and uncontrolled. Finger pointing may be a simple way to clear the accident report up. However, unjustified accusations can be costly and will discourage cooperation with the company's total safety program.

The occurrence of an accident clearly indicates one or more of the following problems:

1. Something has gone wrong in the process, operation, or tasks.
2. Someone has failed to perform a task properly or safely.
3. A hazardous condition exists without adequate safeguards.
4. A process, substance, or defect exists and creates a hazardous potential.

No matter how minor, all accidents and near-miss occurrences should be investigated. Serious accidents arise from the same circumstances as minor instances. In fact, often times, the difference between a minor injury, a near miss, or a fatality is simply a matter of luck.

Immediate on-the-scene accident investigation provides the most accurate and useful information. The longer the delay in examining the accident scene, interviewing people, and establishing the accident facts, the greater the possibility of additional accidents and incomplete investigations. With time, the accident scene changes, witness memories fade, people share accident information, and the initial impressions and observations tend to change.

Accident investigation is not a simple process. It is often difficult to look beyond the immediate incident and uncover the underlying cause.

A major weakness of many accident investigations is the failure to establish and consider the human, situational, and environmental factors.

Some reasons for incomplete or inferior investigations include:

1. Inexperienced and untrained investigators.
2. Reluctance on the part of the investigator to assert authority.
3. Narrow interpretations.
4. Judgmental behavior.
5. Incomplete or erroneous conclusions.
6. Poor interviews.
7. Delays in accident investigation.

ACCIDENT INVESTIGATION SAFETY CONCERNS

The first concern of the accident investigator is the responsible protection of self and others during the accident investigation. In many cases, an accident scene presents a greater danger than a normal work site. For example, structural damage, fire damage, electrical exposures, and environmental factors may still create a direct hazard for the investigator.

It is a wise investigator who remembers that someone was injured at this scene and, therefore, acts cautiously until the causes are known.

Clear responsibilities should be established and delegated for the investigative task. The accident site should be examined and hazards abated before the investigation begins.

HANDLING PHYSICAL EVIDENCE

When physical evidence is mishandled, it renders that evidence useless, making it more difficult to determine the causes of an accident.

The end result of any accident investigation is to prepare for the legal possibilities. Therefore, evidence should be properly identified and secured so that its value will not be lost or impaired.

To be useful in court, evidence must have a proven and unbroken "chain of custody." That means that proper documentation must show where the evidence has been and who has access to that evidence. This prevents uncertainties as to the validity of the evidence.

INVESTIGATIVE TECHNIQUES

Site Documentation Photography

Photography serves as a valuable method of recording conditions that may change during an investigation or shortly thereafter. Photography aids in preparing accident reports and analyzing conditions at the site of an accident.

Photographs of the overall accident scene, wreckage area, and pertinent site conditions should be made prior to any unnecessary rearrangement of the accident scene. These preliminary photographs are helpful in determining the relationships of objects and site conditions prior to investigation.

Before taking photographs, make a determination if the accident scene has been altered. Any items that have been moved or changed should be noted with the circumstances necessitating that change and the persons involved. For example, items may need to be moved for rescue operations, fire-fighting procedures, or other legitimate reasons.

Photographs of accident scenes are broken down into two broad groups—scene-establishing and scene-highlighting photos.

Scene-establishing photos are the preliminary photos of the accident scene that establish locations and the overall relationship of objects. Wide-angle photos from a height or distance to show the overall accident scene fall into this category.

Scene-highlighting photos are those that concentrate on a specific aspect of the accident scene to provide greater detail or evidence. Photographs of broken support columns in a structural collapse would be included in this group.

Photographs in each group should include varying distances (overview, mid-range, and close-up). The use of these three types of photographs clearly establishes relationships and patterns. Photographs should also be taken from several angles. Photographs can often catch and retain what the eye has seen and forgotten. But combinations of camera optics, lighting, and angles can also conceal detail.

Remember the four-cornered approach (front, back, left, and right) for angle photographs. Each photograph should be logged (see Figure 9–1) to record the details of what was taken, when it was taken, and why it was taken, as well as any technical aspects of the photography.

A pocket compass is a useful tool to establish directions from which the photographs were taken.

For close-ups, remember to include a measurement reference such as a tape measure or folding ruler next to the object to be photographed in order to provide an accurate reference scale for the pictures.

In addition to the photography log, the following information should be written or attached to the back of each picture:

1. Accident identifier name or number.
2. Company identifier.
3. Accident scene location.
4. Date (month, day, year).
5. Time of day.
6. Brief description.
7. Sketch reference number.

SKETCHES AND SCENE DRAWINGS

A scene sketch or drawing is a useful tool to document the relationship of objects in the accident scene and to note the location from which photographs were taken.

DATE OF ACCIDENT ____________________
ACCIDENT IDENTIFIER ____________________
LOCATION ____________________
DESCRIPTION ____________________
____
____

#	PHOTO TYPE	TIME	REFERENCE	NOTES
1.				
2.				
3.				
4.				
5.				
6.				
7.				
8.				
9.				
10.				

CAMERA TYPE ____________________ LENS INFO ________
FILM ____________________ ASA ________ ROLL ______ OF ______

Figure 9–1 Accident Scene Photo Log

It is a good idea to mark photo locations by number each time the photographer changes position. This technique will allow for an accurate re-creation of the investigative procedures at a later date.

The following points make scene sketching easier and provide a high degree of accuracy.

1. Always use graph paper to record scene sketches or drawings, and provide a reference value for each square.
2. Orient each area sketched with a north-pointing arrow.
3. Measure diagonal measurements using a script of the same graph paper.
4. Important objects should be located and identified by approximate outline and label.
5. If you have a great number of objects to detail, identify them by numbers, and place the corresponding labels on the back of the sketch page.

6. To establish critical distances, note their locations measured from two fixed reference objects.
7. Note the location and approximate distances of witnesses' vantage points at the accident scene.
8. Note the location from which each group of photographs were taken.
9. Be sure to identify each sketch on the back with the same type of information used on the photographs.

CONDUCTING INTERVIEWS

The basis for conducting a comprehensive interview is to ask those six key questions: who, what, where, when, how, and why.

The "who" questions should answer, Who was injured? Who installed the equipment? Who was responsible for it? The nature of the accident will determine the exact questions you should ask.

The "what" questions should answer, What happened? What did the people do? What equipment or facilities were involved? This line of questions should lead you into actions, events, and physical objects.

The "where" questions should answer, Where was each worker located? Where was the work in progress? Where was the equipment located? The "where" questions have a way of helping you to determine what caused the accident and the conditions that brought that accident about.

Then "when" questions should answer, When did this action happen? What happened next? The answer to the "when" questions should contain more information than simply reading a clock.

Although time may be an important factor in an accident, the relationships within that accident occurrence are even more important. "When" questions often elicit information on the relationship between activities or events.

The "how" questions should provide information on the interaction of people and events. "How" questions refer not only to the actions of the physical environment and equipment but the actions of the injured parties and the witnesses as well.

Answers to "why" questions focus on the unsafe acts, hazardous conditions, and corrective actions of the occurrence. "Why" always leads to more questions. Questions lead to conclusions.

WITNESS CREDIBILITY

The six basic questions—who, what, where, when, how, and why—provide the foundation upon which an investigation develops.

The purpose of investigation is not to establish points that fit a preconceived idea, but to compare points made by a number of witnesses to uncover facts. Witnesses are human and, as such, are subject to normal human failings. They make mistakes, they mislead (intentionally or unintentionally), they withhold information, they exaggerate, and they often interpret what they see.

When taking statements from witnesses, do not assume the information is always valid. No matter how honest witnesses may be, their accounts of the accident will inevitably be colored by individual personality and perspective. Indeed, where you find total agreement from witnesses, you can generally assume that they have already talked the situation over and come to an agreement.

In this case, you job is not only to listen and to record statements, but to discuss and to question to draw out more detail.

When you are speaking to witnesses, remember the tale of the six blind men and the elephant.

> And it came to pass that six blind men on a journey together came upon a large animal. Not knowing what it might be they stopped for a moment in wonder. "Ah," said the first, "it is like a great wall" as he felt the beast's side. "No," stated the second, "it is like a tree," he said as he felt its leg.
>
> "I believe it is made of rope," said the third as he felt its tail. "No," said the fourth, "It is pointed like a spear," as he felt its tusk. The fifth cried, "It's a snake," as he grabbed the trunk. And the sixth listened to the words of his friends and saw a picture in his mind and knew it was an elephant.

Each of them told the truth from their perspective, and yet, it took all of them to get the entire picture. Your job in the investigation is to construct the entire picture, using the pieces obtained from interviews, evidence, and your investigation.

INTERVIEW TECHNIQUES

After obtaining your background material on this accident, the next step is to interview all the involved employees and witnesses. These basic principles will assist you in conducting successful interviews.

Put Them at Ease

First, put the employee at ease. Remind them that the purpose of the interview is solely to prevent a recurrence of the accident, not to find fault. Convince the employee that a joint effort between the two of you will help prevent more serious accidents. Be friendly and understanding.

View the Scene

Conduct the interview at the scene of the accident whenever possible. It will help the employee to explain what happened. However, try to make the interview as private as possible to get his or her true feelings without embarassment. If necessary, view the scene, and then conduct the interview elsewhere.

Listen to the Employee's Story

Ask for the employee's version of the accident. Let him or her tell it as he or she saw it, without interruption. Don't make judgmental remarks which might put the person on the defensive.

Ask Clarifying Questions

Ask any needed questions only after the employee has related his or her comments. Ask open questions that require open responses, rather than yes–no questions.

Close on a Positive Note

Close the interview on a positive note. The best way to close the interview is to thank the person for their time and cooperation and discuss actions that can be taken to prevent future accidents.

Include Other Employees

Interview other employees. The same techniques that were used to interview the employee involved in the accident can be used to interview witnesses to the accident or other involved parties, such as foremen, co-workers, maintenance personnel, or anyone else who can provide additional information about the accident or suggest corrective action.

THE SYSTEM APPROACH TO ACCIDENTS

The factors that make up the system of the work environment are:

1. The people doing the job.
2. The tools and equipment used to do the job.
3. The environment in which the job is done.

These elements combine to complete any task or job. An accident is caused by a flaw in the people, the tools and equipment, or the environment, or often, a combination of the three. An effective accident investigation examines all three elements of the system.

1. Process = Engineering
2. Performance = Training/Supervision
3. Safeguard = Regulations

The Person

Of these three factors, the employees are a critical key. Their actions account for 85 percent or more of all industrial accidents.

Employee accident causation has been cited as a key factor in the following materials:

1. Dupont STOP Program: "Employee actions account for 90 percent of all accidents."
2. OSHA Electrical Safe Work Practice Standard Preamble: "Fifty percent of fatalities are due to employee unsafe acts."
3. Bonnerville Dam Study: "Unsafe acts by employees account for 7 times the number of accidents caused by unsafe site conditions."

There are questions you should ask to guide your investigation of the employee and his or her actions that may have contributed to the accident.

- Was the employee placed on the right job?
- Did he or she have the skills needed to do the job?
- Did he or she have the physical and mental ability to do the job?
- Was the employee properly trained for the job?
- Was he or she experienced in the job?
- Was he or she tired, using medication, drinking, or using drugs?
- Was he or she under emotional stress, worried, or having distracting personal problems?

Tools and Equipment

The tools and equipment we use to make our jobs easier and safer are not without danger. There are many items to consider in investigating the tools and equipment used in a job that might have contributed to an accident.

- Was the machine working properly?
- Was it adjusted correctly?
- Was it the right tool or machine for the job?
- Was it properly guarded, with guards adjusted and working correctly?
- Was the stock or material correct and positioned correctly?
- Was the tool maintained properly?

The Environment

Environmental factors account for 5 percent of all industrial accidents.

The employee and the tools or equipment combine in a work environment to perform a job. Certain factors in the environment should be considered:

- Was the area well lighted? Too hot or cold?
- Was the floor surface in good condition and clean?
- Was the area crowded or congested?
- Was the area noisy, or were vapors, smoke, etc. present, causing distractions?
- Did this noise, smoke, vapors, etc., present a health hazard?

Other Accident Factors

Other factors that contribute to system malfunctions (accidents) often do not fit nicely into one of the three topics as discussed above. For example, consider:

- Time of day.
- Adequate supervision.
- Effects of other people.
- Seasonal or personal considerations.
- Methods of doing the job.

ACCIDENT PREVENTION GOALS

The goal of accident investigation is the prevention of accidents. To accomplish this, however, action must be taken by management (including line supervisors) to eliminate hazardous conditions.

Accident reports should be routed throughout management, reviewed, and channeled for corrective action. These actions may be short or long term or a combination of both.

After corrective action is taken, a program of follow-up and accident analysis is needed to help spot trends and evaluate the effectiveness of the corrective action taken.

Follow-up actions can be summarized by the letters C A L (*C*orrect the condition, *A*nalyze the reasons, *L*earn from our mistakes).

Prepared and Recorded

A statement provides a record in the witnesses' own words. Statements can be handwritten by the people making the statement or recorded and transcribed later.

Statements taken after interviewing a witness should be in a form that is consistent and will be useful as a record later on. Statements should be clear, brief, and complete, concentrating on the accident and issues at hand.

Written Statements

Whenever possible statements should be handwritten by the person making the statement, then signed and witnessed in the investigator's presence.

As an alternative the statement can be written by someone else and signed by the person giving the statement. In these cases a statement like, "I have read the above statement and agree that it is a truthful and accurate account of my knowledge of this incident. I have given this statement freely and openly without persuasion or duress," should appear at the end of the statement just before the person's signature.

Any mistakes made in the written statement should be crossed out, not erased, with the corrections initialed by the signee.

Recorded Statements

Care must be taken in recorded statements to establish the identity of each person's voice who appears on the tape.

Recorded statements should open and close with the following:

"My name is (investigator's name). I am taking this statement on (time, date, and place) from (person's name), regarding (case identifier information).

"(person's name), are you aware that this statement is being recorded, and do I have your permission to do so?"

In recorded statements be careful to clearly establish the accident sequence and details by asking specific questions. For example:

Q: "Mr. Smith the vehicle that was hit on the left side— was that the driver's or passenger's side?"

A: "Driver's side."

Q: "Where was the front of the vehicle pointed when it stopped moving?"

A: "North."

Q: "When you say North, Mr Smith, do you mean towards Main Street or towards Broadway?"

A: "Broadway."

Conduct an interview with each witness to establish what facts the person can attest to. Use the interview to develop the flow of the written or recorded statement you will take. Then prepare specific questions to highlight and clarify points for the statement.

Summary

After each statement is taken, the investigator should write a brief summary of what key points the statement can establish. For example:

"Mr. Smith's statement can establish the direction of travel and final rest position of the accident vehicles and the point and circumstances of impact."

ACCIDENT SCENE EXAMINATION

Sample Taking

There are many reasons for taking samples of materials during an accident investigation. Samples may reveal the cause of a death or injury.

In cases where it is suspected that some mechanical or structural failure contributed to the accident, you may wish to take samples of soil, concrete, or parts of machinery suspected of failure.

Samples of materials taken from the scene may reveal, for example, that insufficiently cured concrete contributed to the collapse of a building or that unstable soil conditions led to a trench or excavation cave-in.

In cases where violations of health standards are the suspected cause of fatalities or casualties, you may find it necessary to take air samples or other samples to check for the presence of impurities or toxic substances.

These samples may show the presence of harmful gases, such as chlorine or carbon monoxide, or of dangerous particulate matter, such as asbestos fibers.

You will find sampling procedures and updated OSHA collecting procedures in the Field Operations Manual and Program Directives for Industrial Hygiene Procedures.

LABORATORY ANALYSIS

Laboratory analysis of samples taken during an accident investigation is usually performed by an independent source. After the samples have been collected and all appropriate paperwork has been completed, samples that are to be chemically analyzed, such as air or dust samples, must be packaged and shipped by certified mail to a certified laboratory of your choice.

OTHER INVESTIGATION CONSIDERATIONS

In addition to the major techniques and approaches to be employed in carrying out an investigation, there are several other, more specific aspects of the accident occurrence with which you should be familiar.

Weather

In some investigations, the weather will be an important factor in determining the possible causes of an accident. It is, therefore, important that you establish exactly what the conditions were at the time of the accident.

For example, an individual could have reported seeing a flash of lightning before being knocked unconscious, yet a check with the weather bureau might confirm that the weather conditions for that day could not have produced lightning. One could conclude that the worker actually had observed an explosion.

Weather conditions can directly affect worker performance or can lead to the use of certain equipment which, in turn, may start the chain of events leading to an accident. For example, flash flooding may cause workers to use a pump before properly checking its safety features.

Clearly, the weather can cause or contribute to accidents. In your investigation, do not omit the weather and its possible role in the accident.

Fire

If an accident involves a fire, it is of the utmost importance that you determine the exact nature of the fire. Was the fire started as a result of the accident? In the case of explosions, did the fire cause the explosion? Did the overloading of an electrical circuit as a result of malfunctioning equipment cause the fire?

Most fire departments have personnel skilled in determining the cause of a fire. However, the evidence may be buried for some time, either literally or figuratively.

Before the analysis and final report are completed, you must know exactly how the fire fits into the overall sequence of events. Do not overlook the fact that the fire and the materials used to put it out can subdue or hide evidence.

Additional information on fire may be found in the *National Fire Codes and National Fire Protection Handbook* published by the National Fire Protection Association. By all means consult these codes, for they can be of help.

Fatalities

Persons killed in an accident may still prove to be one of the most informative sources of evidence at the scene of the accident.

For example, the autopsy report may identify both the cause of death and other unusual conditions or injuries present, though not significant, in the fatality.

Various findings, such as heart attack, foreign gases in the bloodstream, or embedded metal fragments, will play a significant role in your analysis of the accident's chain of events, as well as its cause and contributing factors.

In addition to the coroner's report, the location and position of the person(s) may be significant clues in the investigation.

For example, they may help to determine the relationship between fire and explosion (as was mentioned before) or whether the person(s) was attempting to avoid a worse disaster.

In the case of each fatality, it is essential that you determine as much as possible about the deceased's job responsibilities, operating procedures, skills, and actions.

This information is important in determining the relationship of each individual to the accident. In your investigation, you should always keep in mind that the person could have:

- caused the accident;
- been an unaware bystander; or
- been attempting to prevent the accident or limit its effects.

Though your most important eyewitnesses may have been killed in the accident, information about them from their autopsies may prove valuable.

Records

There are several types of records that may provide a considerable amount of evidence. These include:

- police accident reports
- fire department response reports
- EMS response reports
- media coverage
- court records
- driving records
- national weather report records
- aerial scene photography
- county and municipal repair records
- county and municipal street and plat maps
- medical records
- citations, investigations, reports, and records

THREE TYPES OF ACCIDENT CAUSES

The causes of an accident are actually a combination of simultaneous and sequential circumstances, all of which must have been present for the accident to happen. Any circumstance that contributes to an accident may be spoken of properly as one of the causes of the accident, making it only one circumstance of a combination.

More specifically, a cause of an accident is any behavior, condition, act, or negligence without which the accident would not have happened. Hence, in seeking causes of an accident, remember that an accident may be the result of the interaction among otherwise innocuous conditions and events.

Since you are concerned with all unsafe conditions and practices involved in each accident, you cannot properly talk about "the cause" of an accident. Investigators who come up with a single cause may have conducted an inadequate investigation and analysis.

Although there may be a predominant factor in this accident, to regard it as the only factor involved in the accident is not likely to be either scientific or professional. Such thinking contributes very little to understanding how the accident might have been prevented.

Adverse circumstances are other factors that may be present, without necessarily contributing to a given accident. It is not always easy to determine whether an obviously unfavorable circumstance actually contributed to an accident.

For example, while intoxication is an important factor in some vehicle accidents, there may be times when the driver's intoxication is not the cause of the accident. A sober driver may drive a vehicle into the rear of one operated by a drunken driver who had quite properly stopped his or her vehicle at a traffic signal.

The drunken driver should be arrested for intoxication, but the accident might have occurred even if he or she had been completely sober. Drunkenness did not necessarily cause the accident, and simply finding a traffic violation (driving while intoxicated) might not be considered a complete investigation to determine causes of the accident.

In order to get away from the idea that most accidents have a single cause, there is need for specific guidance in defining and identifying accident causes.

Such guidance will help you to improve your investigation and ensure better analysis, and set the stage for more appropriate recommendations for corrective action. For these purposes, accident causes can be classified into three easily recognizable levels: direct, indirect, and contributing.

Every accident is the result of at least one direct and one indirect cause, and some accidents have many of each. These are briefly described below.

Direct Causes

Direct causes are unsafe acts, doing something or failing to do something required by law, and taking action that can be tied directly to the accident or incident. These include actions that are violations, improper, hazardous, unexpected, or illegal. For example:

1. Operating at an unsafe or improper speed.
2. Unsafe employee acts.
3. Unsafe site conditions.

4. Inattention, impairment, or improper actions or responses.
5. Deliberate violations of safety regulations and practices.
6. Failure to abate a hazard, or faulty or improper actions to correct a hazard.

Indirect Causes

Indirect causes include irregular or unusual conditions that explain why the person, equipment, or operation contributed to an accident occurrence. Indirect causes are connected to the accident through the direct causes.

Examples of indirect causes include:

1. Defects in equipment, materials, tools, or structures that contributed to the occurrence.
2. Contributing weather conditions or unusual weather conditions.
3. Problems with visibility, control, safety-procedure breakdown, or terrain problems.
4. Contributing conditions in employees, i.e., poor eyesight, lack of training or knowledge, exhaustion, stress, and physical or emotional conditions.

Contributing Causes

Negligence or conditions not directly responsible for an accident but necessary for the accident to have occurred are considered contributing causes. These include:

1. Inadequate codes or standards.
2. Lack of effective company safety policy.
3. Inadequate supervision.
4. Faulty design.
5. Inadequate maintenance.
6. Poor enforcement.
7. Substandard craft work.

Most accidents and incidents involve a number of causes, all interrelated. The most common failing in accident investigation is to look for the easy answer.

Unless we understand the precise causes and contributing factors involved in an accident, we are powerless to take effective corrective actions.

FAILURE MODES

Causative factors can be grouped into several broad classifications, called "failure modes." An understanding of these types of classifications can be useful in directing the scope of your investigation.

Human Failure

Physical, psychological, and pathological limitations are included in this class. Human factors may be underlying and hidden. This makes them difficult to uncover. For example, failure to follow correct safety procedures, illness, distractions, stress, impairment, and deliberate actions may be involved.

Even when the person is aware of his or her own contribution to an accident occurrence, that person is likely to blot out responsibility or use excuses to cover oneself from blame.

It is often possible to determine that human error was involved but not the underlying causes and contributing factors that created that human error.

The Stanford University School of Engineering conducted a survey of worker-safety attitudes. They concluded that some workers accept risk taking as a normal part of the job. These workers tend to trust luck to keep them safe and are more prone to accident occurrences.

Technical Failure

This class results from the use of incomplete, inadequate, or erroneous data—through the omission of critical data or the use of improper data, operating instructions, or faulty designs.

Organizational Failure

This exists when management deficiencies contributed to the occurrence, or when improper management response to a situation worsened its occurrence. This is often due to inadequate planning, supervision, authority in a crisis, work policy, and procedures or an improper safety effort or execution.

Material Failure

The physical or chemical breakdown of any part of the structure or components, the use of improper or substandard materials, or using materials or components improperly contribute to this class of failure.

Natural Phenomena

Acts of God, freaks of nature, or the result of severe weather, war, or other disaster-producing agents are involved here. This does not apply to any failure to take proper and adequate precautions against predictable contingencies.

Motor Vehicle Accidents

According to the National Safety Council's Accident Facts for 1988, there were over 4,000 work-related motor vehicle fatalities in 1987 and 1,600,000 work-related disabling motor vehicle accidents.

Motor vehicle accidents cost 64.7 billion dollars in 1987. This includes lost wages, medical costs, insurance administration costs, and property damage. This did not include the cost to public agencies like police and fire departments, court costs, and indirect accident costs.

Generally, the driver is the only representative of the company on the scene at the time of the accident. Each driver should be trained to take the following steps whenever possible.

1. *Protect the scene.* The driver should immediately place warning signals and devices to prevent further injury and to control the normal flow of traffic safely.
2. *Injury assistance.* If possible, the driver should request assistance for any injured parties.
3. *Report the accident.* The driver should have a telephone number and procedure to report the accident to the company. A prepared telephone checklist is invaluable in re-

cording the information reported on a motor vehicle accident.

4. *Ensure vehicle records are in order.* Each vehicle should be equipped with an accident report form carried in the glove compartment. All accident-reporting forms, reporting-procedure memos, company insurance and identification information and vehicle documents should be contained in this package.

   Many companies place a notification sticker inside the vehicle to indicate who should be notified in the event the driver is incapacitated.
5. *Obtain information.* Whenever possible, the driver should gather preliminary information at the accident scene, noting the identity of the vehicles and persons involved as well as the names, addresses, and phone numbers of witnesses. Preliminary accident report forms should be part of the vehicle accident report kit.

MOTOR VEHICLE SCENE INVESTIGATION

When the accident investigator responds to the scene of a motor vehicle accident, he or she should take the following steps:

1. Park your vehicle so that it will not obstruct traffic or contribute to another accident.
2. Ensure that warning devices have been placed properly.
3. Provide identification to the police at the scene.
4. Interview company personnel or the driver involved.
5. Photograph the accident scene.
6. Canvas the bystanders for possible witnesses or assistance if necessary.
7. Take steps to protect company property from further damage, theft, or vandalism.
8. Begin a complete investigation.

RECORDING ACCIDENT FACTS AND DETAILS

Each accident presents its own unique problems. The following information should be used as a guideline for the details the investigator should obtain.

1. Time. Record the exact day, date, and hour of the accident. If the vehicle is equipped with a tachograph, this will be charted.
2. Accident location. Recording specific information to clearly establish the accident scene is necessary.

   Take measurements from at least two fixed reference points, and be sure all road areas are identified. Use utility pole designation numbers, mile markers, and other standard reference points whenever possible.
3. Accident involvement. Specific details about the make and type of each vehicle involved is important. Also, identify each person by name, address, telephone number, and position in the vehicle. Identify all fixed objects or pedestrians.
4. Accident type. Classify and record the pertinent details that accurately describe the accident occurrence.
5. Driver information. Obtain the identification address, phone number, and driver's license information for each driver. Also, obtain complete insurance information for each vehicle or location of property damage.
6. Injuries. List the name, address, age, sex, vehicle position, and injuries of each injured person.
7. Vehicle damage. Describe and classify the damage to each vehicle, noting all damaged areas, marks, and debris locations. Also, describe damage to cargo, other company property, and all other personal and public property.
8. Witness canvas. Witnesses need to be identified as soon as possible. List their names, addresses, and phone numbers. In addition, the identity of all responding units and officers (police, fire, and ambulance) should be recorded.

   A record should be made of the license number of all vehicles parked at the accident area, and the names and numbers of all buildings overlooking the accident scene should be noted.
9. Vehicle movement. Determination of each vehicle's movement prior to the accident occurrence should be made. As accurately as possible, reconstruct and diagram the movements after interviewing the drivers, passengers, and witnesses.

10. Contributing factors. Investigators should examine and record the mechanical conditions of the vehicles, noting any statements made due to mechanical failure.
11. Weather. Weather conditions at the time of the accident should be recorded. The accident report should include indication of whether any weather conditions could contribute to the accident, for example, rainy, slippery roads, sun glare, windblown debris, etc.
12. Road conditions. The roadway conditions should be noted, and accurate measurements of the roadway at the accident scene should be made.

    Pictures should be taken of all road-control devices (traffic signs, traffic signals, etc.).

    The investigator should attempt to determine whether improper or inoperative traffic-control devices may have contributed to this occurrence.
13. Accident diagram. An accident diagram should be complete and show at a glance the circumstances of the accident occurrence. The use of plotting devices, such as the Northwestern University Traffic Investigator Template, can assist in creating accurate diagrams.

    Accident diagrams should contain the following information:
    a. width of the roadway
    b. conditions of roadway
    c. width, number, and marking of each traffic lane
    d. width, conditions, and drop-off measurements of the road shoulder
    e. point of impact indication, with debris locations noted
    f. dimensions, conditions, and markings of each street, roadway, alley, etc.
    g. locations and directions of each vehicle traveled prior to and after impact
    h. location reference markers and distance measurements to critical accident-scene components
    i. debris location, skid marks, and other physical evidence
    j. location of all traffic signals, signs, and devices
    k. notations of any device that obstructs driver vision

l. location of all fixed objects related to the accident scene with related measurements
   Remember that, in taking measurements, triangulate from the object measured to two known points of reference and between the two reference points should be used.

14. Photographs. All photographs and the photo log should be attached to the accident report. A sample photo log is presented in Figure 9–1. Remember to record identification information on the back of each photograph.

    Accident scenes should be photographed from a distance, middle, and close ranges. All critical information should be documented in front, back, and both sides (four-cornered method).

    When extreme close-ups are taken, establishing shots showing their relationship to the larger part of the accident scene is recommended.

15. Ancillary materials. Driver statements, witness statement, and witness lists along with all pertinent outside reports (police, fire, and ambulance) should be listed or attached to the report or provided in a supplemental report.

Chapter 10

Drugs in the Workplace

DRUGS AND ALCOHOL—LEGAL ISSUE

The decision for a company to develop a substance abuse (drug and/or alcohol) program involves complex legal issues. Nothing within this chapter is intended as, or implied to be, legal advice. This chapter discusses the issues in general surrounding the development and implementation of company drug and alcohol policies and programs. The advice of legal counsel is necessary to address the specific legal aspects of your company's situation and needs.

DRUG USE AND ABUSE

We are all familiar with the problem of drug use in our society. No one would deny that the problem exists and that it is extensive and pervasive. We are reminded of the drug issue, with its concomitant crime problems, every time we turn on the television or read a newspaper.

Yet, we often fail to recognize that the drug issue that contaminates every level of society also impacts our places of business. Many of us turn a blind eye to workplace drug problems or even fail to admit their existence.

Perhaps, we think that the drug problem in our society does not spill over into the workplace—that somehow the workplace has some special immunity from drug use.

Unfortunately, the drug user does not suddenly become a nonuser when he or she steps on your work site. If a person has such little regard for their own home and life that he or she will abuse drugs within them, then that behavior will carry over into the workplace.

It is the impact of drug use in our workplaces that must be dealt with. The effects that drug abuse has on the safety of all other workers on site is the same, whether the drug is used within the confines of the workplace or when the drug-impaired worker arrives at work.

Because no level of drug impairment in the workplace can be safely tolerated, the only acceptable level of drug and alcohol use in our workplaces is zero. Many workplaces are adopting a zero-tolerance policy for drug use.

DRUG ABUSE IN THE WORKPLACE

Scope of the Problem

Statistics from recent congressional hearings on drug abuse point out the costs and effects of drug use in the American workplacc:

1. It costs American businesses approximately $6 billion a year for problems and programs related to workplace drug abuse. This averages out to be between $3,000 and $6,000 per employee.
2. Loss productivity costs for U.S. industry-related drug abuse are estimated to be $30.8 billion each year.
3. American industries lost $81 billion in 1984 due to accidents. An unknown percentage of that loss is related to drug abuse.
4. Construction industry experts estimate that 20 percent to 40 percent of the work force abuse substances on the job site.
5. The National Cocaine Helpline offers counseling and information to people with drug problems. Since its inception, it has averaged 1,000 calls per day. Unlike the stereotypical profile of the addict as a lower-class street person, the average caller is a white, 30-year-old, employed male,

earning more than $25,000 per year with an average education of 14 years.

6. Recent reports from the Drug Enforcement Agency indicate that the world production of illegal drugs has risen dramatically. More and more drugs are being dumped into the American market.

THE CONSTRUCTION INDUSTRY INSTITUTE

In September 1987, the Construction Industry Institute held a national forum on workplace drug abuse in Washington, D.C. According to a survey done by Dr. Bill Maloney:

1. Twenty-seven percent of the construction companies surveyed said there was a serious workplace drug problem.
2. Forty-three percent indicated they had a drug program in their company.
3. Only 12 percent of the construction companies surveyed tested workers for drug use.

This reluctance to engage in employee drug-testing programs does not come from a lack of understanding of the serious workplace problems associated with drug abuse. We know that drugs, including alcohol, negatively affect our workplaces every day. The absence of widespread testing programs stems from the quagmire of potential legal, contractual, and economic problems surrounding the drug-testing issue.

The recent increase in companies developing drug policies and programs is based, in part, on the growing national awareness of and educational efforts on substance abuse.

This, coupled with the federal government's "war on drugs," yields more employer acknowledgment of the drug problem and emerging court precedents are beginning to give companies more latitude in their testing programs and efforts.

The Construction Industry Institute (CII) survey also cited the characteristics of a typical drug abuser. That person:

1. has three times the absenteeism rate of non-drug-abusing employees
2. uses three times the sick benefits
3. files five times as many worker's compensation claims

4. has 3.6 times the number of accidents
5. is only 67 percent as productive as nonabusers.

Alcohol use by construction workers is traditionally more acceptable and viewed, somehow, as not as serious a problem as drugs. However, alcohol is also a drug, and for alcohol abusers the negative characteristics remain the same.

If we use the minimum percentage of drug abusers in the construction industry quoted in various studies, multiplied by the total construction-industry employment, then there are over 600,000 drug abusers in the construction workplace on any given day.

The impact of drug abusers on construction costs is reflected in the following CII statistics:

1. Drug use increases overall construction costs by 8.4 percent.
2. Drug use increases health-care costs by 16.4 percent.
3. Drug use increases worker's compensation costs by 17.6 percent, and it increases liability costs by 14.4 percent.

These figures are alarming, but they do not consider the total costs involved, for example, inflation and increases in construction-insurance costs.

A COMPANY DRUG PROGRAM

Instituting a company drug- and alcohol-abuse program is a direct way for a company to deal with and control the costs associated with the impact of drug and alcohol abuse in the workplace. The controversial issue is whether or not to test for drug use.

WHY COMPANIES ARE TESTING

Companies test in response to an actual or perceived drug-use problem within the workplace. They test as a precautionary measure to reduce accidents and injuries and to reduce medical and insurance claim costs.

Companies test to fulfill their obligations under the OSHA Act, to provide a safe workplace, and because drug abuse presents a clear danger to their most valuable resource—their employees.

Companies test because it is good business. They test to reduce costs and absenteeism and to increase productivity. Companies also test because of contractual requirements and new federal laws such as the Drug Free Workplace Act of 1988.

IMPLEMENTING A COMPANY SUBSTANCE ABUSE POLICY

Table 10–1 outlines the steps in implementing a company substance-abuse policy. One of the first considerations for your company is the decision to test or not to test.

UNION AND OPEN-SHOP CONSIDERATIONS

Companies with union contracts must consider their statutory and contractual rights and obligations, under their collective bargaining agreements. Many union contracts place limitations

Table 10–1 Steps in Implementing a Company Substance-Abuse Policy

Implementation varies from union to open-shop contractors. In union shop situations, the National Labor Relations Board has indicated that drug testing and disciplinary action programs are mandatory subjects of collective bargaining. Employers must first attempt good-faith bargaining until agreement or impasse before implementing a company drug program. Open-shop contractors have much more latitude in their implementation.

1. Issue the company a formal written drug and alcohol policy.
2. Ensure that all supervisory personnel are trained in the program's implementation and administration.
3. Provide employee orientation training in the reasons for and specifics of the company drug policy and any disciplinary actions for violations.
4. Monitor all occurrences under the scope of the policy to ensure consistency in application.
5. Evaluate the program elements periodically and revise the program as necessary.

on the employer's right to establish and enforce work or disciplinary rules.

The National Labor Relations Board's (NLRB) decision on drug testing restricts a unionized employer's right to arbitrarily change work policies.

The General Counsel's Office of the National Labor Relations Board issued a memorandum, on September 8, 1987, to the regional offices, which provides guidelines for pending and future cases concerning drug or alcohol testing of employees.

According to the NLRB General Counsel, drug or alcohol testing for current employees and job applicants is a mandatory subject of collective bargaining. Any employer subject to NLRB authority who seeks to implement drug or alcohol testing of employees must notify the union of its intent to initiate such testing. Employers, upon request, must bargain to an agreement or "good faith" impasse before implementing any such program.

COMPANY POLICY STATEMENTS

Establishing a company policy starts with the development of a clear and definitive company statement on drug abuse.

The policy should be in writing and distributed to all employees. Employers today are lucky in that they can learn from the past efforts of other companies.

Policies in use today that have withstood the challenges and opposition to drug testing are valuable resources to use in creating your company policy.

The following key elements are generic to many construction company drug and alcohol policies currently in use and may be useful for consideration in drafting your company's program:

1. A statement of the need for a drug policy such as:

   "Employees are our company's most valuable resource, and for that reason, their health and safety is of paramount concern.

   The abuse of legal and illegal drugs on or off the job can result in increased accidents and injuries. Therefore, to protect our employees, we are instituting the following."

2. A statement outlining the company's "Zero Tolerance Rule." This rule mandates that,

   "The company will accept no level of drug or alcohol use above zero."
3. A statement defining the prohibitions set by the company policy.

   "The illegal use, sale, distribution, or possession of narcotics, drugs, controlled substances, or the misuse of legal drugs including alcohol while on the job site or on company property is prohibited."
4. A statement defining the company's action for violations of company policy.

   "Any violations of this policy (may) result in disciplinary actions up to and including discharge. The sale of illegal narcotics, drugs, or controlled substances off duty and off company premises (may) also result in discharge or disciplinary action."

Prohibition statements within the company policy may be broad ranging and include some or all of the following language:

5. The illegal use of drugs off duty and off company premises is not acceptable. Such use can affect on-the-job performance, job safety, and the confidence of the public and the government in the company's ability to meet its responsibilities. Such use may result in actions up to and including discharge.
6. Alcohol use and possession is prohibited from company property and operations. The use of alcohol on or off the job that adversely affects an employee's job performance or the public and/or regulatory perception of the company is not acceptable and may result in discharge.
7. The legal use of controlled substances prescribed by a licensed physician is not prohibited. Employees in selected positions, designated by the company, are required to make such use known to an appropriate company representative. For safety considerations, the company reserves the right to restrict the employee's work operations or to temporarily reassign employees to other duties.

8. All company work areas, facilities, sites, and equipment are subject to routine and unannounced searches by company representatives for contraband and prohibited substances.
9. All personnel entering or leaving site, and any company or personal possessions or materials carried onto or off work areas, are subject to search.
10. Any contraband or prohibited materials found on company property are subject to confiscation.
11. Law enforcement officials will be notified whenever suspected illegal drugs or drug paraphernalia are found.
12. All vehicles entering or leaving company property or parked on company property or work sites are subject to search at any time.
13. Any employee refusing to comply with a request for a search may be subject to disciplinary action up to and including immediate termination.
14. Whenever possible, the company will provide information to and/or assist employees in overcoming drug, alcohol, and other problems that may adversely affect the employee's job performance.
15. As part of an overall substance-abuse program, the company may utilize drug-testing procedures, including preemployment, for cause and random tests.
16. Employees failing to comply with a company request to submit to drug testing may be subject to disciplinary actions up to and including discharge.

DRUG-TESTING CRITERIA

Before a company institutes a drug-testing program, it should document and demonstrate the need for drug testing within the company. This is usually done by quoting the national statistics that suggest a relationship between negative job performance and substance abuse.

Employers should also cite their responsibility, under OSHA statutes, to provide a safe work environment for all employees.

The company should develop a specific company substance-abuse policy and program, in consultation with all the areas of the company that may be affected. By involving the employees and management in development of the drug policy, a company can avoid the "They're out to get me syndrome" in the employees and increase the program's acceptance.

In your company situation, consider involving your union representatives, occupational health and safety personnel, security staff, personnel managers, legal advisers, top management, and employee representatives in the program development. This can easily be done by having a committee review program drafts and provide comments and feedback.

After developing a specific policy, you will need to modify work and union contracts to reflect the company's new substance-abuse policy where necessary.

As soon as possible, notify all your employees of the new policy. Tell them in advance the specifics of the policy and the penalties that will be imposed for violations. Set a reasonable time for program start-up.

Consider establishing an initial amnesty period within the program. During this period, employees who voluntarily reveal a substance-abuse problem may retain employment after successfully completing a rehabilitation program. Amnesty programs are good for morale and place the employer in a positive light against claims for discrimination against affected employees.

Once your program is established, follow through, don't let your substance-abuse program become a "paper" policy. Your policy must be reviewed and updated to fit your current needs.

Your policy should be reviewed on at least an annual basis to see if any changes are necessary.

If your company's drug program will include testing, be sure to establish a clear criteria for all aspects of drug testing. Generic drug testing guidelines are provided in Table 10–2.

Establish a formalized mechanism to notify the employee of any positive test results. Your policy should provide employees with an opportunity to appeal disciplinary actions taken on the basis of positive tests results.

Keep tests results confidential. Do not take any actions on a positive screen test result until its accuracy has been verified by a confirmation test.

Employees are a valuable resource. Consider setting up an Employee Assistance Program (EAP). These programs can be set

Table 10–2 Guidelines for Drug-Testing Program Criteria

1. Use an EMIT screening with G-MASS SPEC confirmation for all positive tests.
2. Utilize a written chain of custody for all samples.
3. Provide for a split sample policy with the employee. This gives the employee the right to conduct an independent test from the same sample. These split tests are conducted at the employee's expense.
4. Use a coding system for all test samples, without test sample names. This way, the laboratory has no knowledge of the test subject's identity.
5. Isolate drug-test results from all other employee records and files to maintain their confidentiality.
6. Use a certified laboratory; consider conducting independent blind tests of the laboratory at periodic intervals.

Table 10–3 Employee Assistance Program Guidelines

When considering an established rehabilitative or counseling program, use the following questions:

1. Who owns and manages the EAP?
2. How do employees avail themselves of these services?
3. What services are offered?
4. How are they paid?

There are two basic types of EAP programs—internal and contractor programs.

Internal programs are generally found in larger companies. They are staffed by specialists employed by the sponsoring organization, employer, or union. The staff assesses the employee's needs and recommends specific treatment and counseling programs.

Contractors are EAP service providers or service centers selling their services to employers. Contractors may also run an employer's internal EAP program.

Table 10–4 Steps to a Successful EAP Program

1. Develop written EAP policies and procedures.
2. Ensure that the EAP remains separate from the grievance/arbitration procedures.
3. Insist on top management support and involvement in all EAP referrals, publicity, and employee information programs.
4. Train supervisors, managers, etc., to discuss EAP information with employees to encourage the employee with a drug problem to seek help.
5. Provide employee training, orientation, and information on the EAP program and the company drug policy and procedures.
6. Monitor the EAP program, its success ratio, and its cost factors in order to consider changes where necessary.

up within the company, or an outside counseling or treatment service can be used. Tables 10–3 and 10–4 provide assistance in establishing an EAP. Most medical insurance coverage will pay for a substantial part of the employee EAP costs.

However, nothing in the current drug regulations and legislation requires an employer to have or to pay for EAPs.

LABORATORY CRITERIA

Quality drug testing begins with the use of a qualified and certified laboratory. When choosing a lab to handle your program testing needs, consider the following questions:

1. What is the lab's testing volume and experience?
2. What methods of screening and confirmation are used?
3. Which drugs is the lab capable of detecting?
4. What kind of quality control program does the lab have, and is it participating in any outside proficiency testing programs? Is the lab currently certified and by whom?
5. What chain of custody procedures and/or documentation does the lab use?
6. What kind of sample handling assistance does the lab provide? Do they take samples or provide sampling supplies?

7. What provisions does the lab have for sample retention? How long do they retain samples and at what cost?
8. What kind of support will the lab provide if legal action is brought against the employer? Will it provide testimony on the lab's qualifications, test accuracy, and lab procedures?
9. What is the lab's turnaround time for sample results?
10. How does the lab report results?
11. How good is the lab's security?
12. What does the lab charge, and what volume discounts are available?

The more information you have on the laboratories, the better choice you can make. Be sure to check with a number of labs in your area to determine costs and quality of services. The authors' research on cost showed that prices can vary as much as 200 percent for the same test procedure within the same geographic area.

When initially interviewing laboratories, ask them to send you a price list and information packet on their personnel, procedures, and qualifications.

Steps are currently being taken to regulate all laboratories engaged in drug-abuse testing. Labs providing testing on an interstate basis are regulated by the Health Care Financing Administration (HCFA), under the authority of the Clinical Laboratory Improvement Act (CLIA) of 1967.

The lab you use should be under the direction of a toxicologist with a title of Diplomate from the American Board of Toxicologists or the American Board of Forensic Toxicologists.

Take the time to examine the lab qualifications and visit the lab before making your final decision.

TESTING ACCURACY

Opponents of drug-testing programs like to make reference to the "inaccuracy of drug testing." They also decry the potential fate of the poor person falsely accused of drug use based on those test results.

Usually, they will cite the Centers for Disease Control (CDC) tests to support their claims. In the early 1970s, the Centers for Disease Control conducted blind sample tests of 27 laboratories across the United States.

They found an alarming inaccuracy rate (up to 67 percent), for "false negative tests result." In a "false negative test," a drug present is not detected.

Opponents of drug testing still use the CDC results today to indicate that all drug-testing results are unreliable. This is no longer a valid conclusion for the following reasons:

1. The drug test methodology accepted by the courts today were not in use at the time the CDC tests were conducted.
2. Testing methodology in use today calls for a broad screen test, to separate the negative from the potentially positive samples. A sample is not reported as a positive test until a confirmation test of a highly specific nature is conducted. Even then, the specific drug detected must be present at an establish measured level in the sample for the test to be labeled positive.

   This testing criteria usually involves an "EMIT" (Enzyme Multiplied Immunoassay Technology)-type screen test with a "G-MASS Spec"(Gas Chromatograph Mass Spectrometer) confirmation. These tests are sufficiently accurate to be considered a definitive test under the court-accepted rules of evidence.
3. A "false negative rate" will not subject an employee to an unjustified charge of drug use because, simply stated, a "false negative result" did not detect a drug that was in fact present in the sample.

   This condition is exactly opposite the concerns implied by drug-testing opponents, which is labeled a "false positive test result."

   In a "false positive result," a drug is mistakenly identified in a sample when no drug is present.

   It is important to note that the CDC tests indicated a high "false negative" and not a high "false positive" result. The laboratories in the CDC tests failed to detect a drug present in the blind sample submitted. Under that condition, the only ones to suffer by a high rate of "false negative" test results are the employers.

No matter how accurate a testing method is, there is always a rare chance of an inaccurate or incorrect test result.

However, good laboratories build checks and balances into their systems to catch these rare analytical errors before the results are reported. This is why a screen and conformation test are now considered standard test criteria for any positive result.

Drug Tests and Impairment

Current drug tests do not measure a direct level of employee impairment. Drug tests simply measure the presence of a metabolite associated with a particular drug's use. It is important that employers clearly understand this fact. Currently, positive test results do not measure impairment levels.

Statements made about an employee's impairment, or drug policies seeking to address the question of impairment, often open the employer to lawsuits.

The "zero" tolerance doctrine of workplace drug use avoids the pitfalls in the impairment issue by flatly stating that no level of drug use is acceptable in the workplace.

TESTING METHODOLOGIES

Employers currently instituting a drug-testing policy or revising their company's testing program have the advantage of choosing procedures that have already withstood the court's scrutiny.

By utilizing an accepted screen test with a confirmation test for all positive screen results, employers can reduce their potential for costly litigation based on inaccurate test-result claims.

Key points to keep in mind in choosing testing procedures include:

1. Treat all positive test samples as potential evidence for a court proceeding.
2. Make sure you have a written chain of custody for test samples. A written chain of custody is a documented record of a test sample possession from the moment the specimen is collected through its transport, analysis, and submission in court.

There are four types of common laboratory tests for drugs used today:

1. Immunoassay
2. Thin Layer Chromatography

3. Gas Chromatography
4. Gas Chromatography/Mass Spectroscopy

Immunoassay: This test uses antibodies that are highly specific to isolate drug metabolites from urine. These metabolites or by-products are then labeled for further measurement and quantified by highly sensitive equipment.

EMIT (Enzyme Multiplied Immunoassay Technology): EMIT is considered to be currently the most widely accepted screening methodology for drug abuse because:

a. It is reliable. Test materials are commercially produced and subjected to stringent quality-control procedures.
b. EMIT can be automated by the use of Rapid Spectrophotometer equipment that helps reduce potential human errors and keeps the drug screen costs in a price range that most companies can afford.
c. EMIT results are objective, with tests results derived by a mathematical formula that uses the absorbance of light as a measurement base.

Thin Layer Chromatography (TLC): TLC requires the technologist to take a visual reading of plates that have been specially treated to physically separate the compounds in the urine sample.

A drop of urine is placed at the base of the plate. The compounds in the urine then migrate at different rates up the plate. When a fixative and color developer are applied to the plate, specific colored spots will appear if drugs or drug metabolites are present. The spots are then classified by their color and position on the plate.

TLC is widely used because:

a. Tests can be used as an economical, reliable, broad spectrum screen test for the presence of a drug.
b. Results can be preserved with a color photograph for comparison or evidence.

TLC has some drawbacks for employer drug test use, including the following:

a. It is a subjective test result. A technician must make a determination based on an individual assessment of the position of the spots on the plate.

b. This test requires years of practice for a technologist to become proficient at interpreting the results correctly.

GC/MS (Gas Chromatography/Mass Spectroscopy). GC/MS procedures have been labeled as a chemical fingerprint test that identifies a substance based on its unique molecular composition.

The high accuracy rating for this test method comes from the fact that the method provides direct retention information via a graph printout of the sample, as well as evidence of the specific molecular structure of the drugs in the sample.

NEED FOR CONFIRMATION TESTING

All drug-screening methodologies have a certainty of error in a small percentage of samples analyzed. This is due to interference with test results by certain substances and human or equipment errors.

Therefore, all specimens presumed to be positive by one test should be confirmed by an alternate test method before being reported as positive. Experts agree that a single testing procedure is not considered defensible against legal challenges of their accuracy.

The court systems have not accepted single-method drug testing under the rule for admissible evidence.

Before taking any action on a positive screen drug test, employers should obtain the results of a secondary or confirmation test. This two-test method protects the employer from successful challenges to the accuracy of the drug-testing methods they used.

DRUG-TEST DETECTION LIMITS

The detectable limits of currently used drug-test methods are extremely broad. Very small quantities of a drug can be detected in a test sample.

In order to eliminate possible inaccurate test results, identify cross-sensitivity reactions, and confirm direct use of a drug, positive test-reporting limits are established by the employer.

A positive test limit is an arbitrary value, below which a test is not considered to be positive for that particular drug. Companies may set that positive reporting limit at any number they

Table 10–5 Suggested Positive Test Limits for Common Drugs

75	100	300	750
		Nanograms per Milliliter (ng/Ml)	
PCP	Cannabis	Amphetamines Barbiturates Benzodiazepines Cocaine Opiates Methadone	Methamphetamine

Note: These limits set the level at which a drug test is considered positive and is court-accepted. They are the recommended positive test limits established by the National Institute of Drug Abuse (NIDA).

choose. There are, however, a set of positive test values recommended by the National Institute of Drug Abuse (NIDA), widely in use by employers and accepted by the courts. (See Table 10–5.)

Table 10–6 Guidelines for Drug Detection Time in the Body

Drug	Time
Amphetamines	48 hours
Barbiturates (short acting)	24 hours
Barbiturates (long acting)	2 to 3 weeks
Benzodiazepines (for tranquilizers like Valium)	3 days
Cocaine	2 to 4 days
Canabinoids (moderate use)	5 days
Canabinoids (heavy use)	10 days
Canabinoids (chronic use)	20 or more days
Methadone	3 days
Methaqualone	2 weeks
Propoxphene	6 to 48 hours
Phencyclidine	8 days

DRUG RETENTION IN THE BODY

How long a drug remains detectable in the body after ingestion depends on many factors, including:

- the dose and frequency of drug use;
- the person's physical condition, health, and fluid intake; and
- the type and sensitivity of the test method utilized.

As a general guideline or rule of thumb, the detection times listed in Table 10–6 can be used.

BLOOD TESTS

Because of the controversy surrounding and intrusiveness involved in blood testing, it is not widely used in company drug-testing programs except in postaccident cases.

The taking of blood in most cases without consent has been seen as an unreasonable search and seizure by the courts.

Therefore, the use of blood tests require the consent of the employee. Such consent can be given on an individual basis or as a blanket consent form as part of a prehire package. The collection of samples should be conducted under the direction of a licensed physician at a laboratory or medical facility.

RECORD KEEPING AND CONFIDENTIALITY

Laboratories seldom provide the forensic quality of testing procedures required to prevail in a challenge of your company's drug-screening program. When choosing a laboratory to handle your drug testing, pay particular attention to their methods of record keeping.

For your protection, your company drug-testing policy should require laboratories to:

1. Maintain extensive written records of testing procedures and sample handling, including:
   a. logs of all specimens,
   b. external and internal chain of custody,
   c. logs of all visitors,
   d. logs of security and maintenance procedures for all controlled substances,
   e. a detailed record of test results and reports.

2. File and maintain all laboratory proficiency results, quality-control sample records, and personnel files associated with your samples.
3. Maintain positive tests specimens in a frozen state for a minimum of one year. Additional storage is required when legal challenge occurs. In these cases, the sample should be maintained until the case is completed.
4. Maintain a high degree of security at the laboratory to prevent any tampering with your company's specimens or records.
5. Accept the use of your company's encoded samples (sample I.D. numbers, not names) to maintain a high degree of record confidentially. The master list linking the code number to the particular employee's name is maintained at and secured by your company. Therefore, results reported to your company cannot be intercepted or accidentally or inadvertently released.

SUPERVISORY TRAINING

A company drug program requires that all supervisory personnel be trained for the role they will perform within the program structure. Supervisors will need training in the following areas:

1. The specifics of the company's drug program.
2. How to recognize drugs, drug paraphernalia, and drug-related signs, symptoms, and problems.
3. The supervisor's role in implementing and administrating the company's drug program.
4. Dealing with employee information requests, employee complaints, and problems.

Table 10–7 deals with general signs and symptoms associated with drug use.

DRUG TESTING ISSUES

The use of drug screening for job applicants and current employees is rapidly increasing in American industry. Opponents of employee drug testing argue that:

1. Drug tests violate the Fourth Amendment prohibition against unreasonable search and seizure.

Table 10–7 Signs and Symptoms of Drug Use

1. Bloodshot or watery eyes.
2. Either very enlarged or very small pupils.
3. Runny nose or sores around the nostrils.
4. Wearing sunglasses indoors and in all weather.
5. Excessive perspiration.
6. Sudden and unexplained weight loss.
7. Sudden worsening of the complexion.
8. Extreme mood swings or changes.
9. Irritable, inappropriate, or unpredictable behavior.
10. Frequent absenteeism.
11. Excessive use of sick leave.
12. Frequent lateness to work.
13. Early departures from work.
14. Deteriorating production quality and/or quantity.
15. Frequent accidents and/or injuries.

Note: The presence of one or all of the above symptoms is not conclusive evidence of a substance problem. No disciplinary action should be taken or accusations made until conclusive testing results are available.

2. The tests results are often inaccurate, depriving the employee of due process.
3. A positive test indicates only the presence of certain quantities of drug residue; it is not evidence that an individual is impaired in job performance.
4. There is potential for abuse of the information revealed through the test by the employer.
5. Employers do not always enforce drug-use regulations uniformly.

In response to these arguments, let us consider the following.

CONSTITUTIONAL ISSUES

Privacy Rights

The United States Supreme Court has found that the right to privacy implied in the Constitution protects only against governmental intrusions (federal, state, and local agencies). Public employees, working for a governmental entity, are bound by Fourth Amendment rights. This freedom from unreasonable search and seizure does not apply to private employers.

In addressing the drug-testing issues, the courts have balanced the need for conducting a search for drugs and the invasion of privacy questions, with considerations of the scope of the intrusion, the manner in which the test were conducted, and the justification for testing.

In relationship to private employers, a search is considered unreasonable by the courts only if it extends into some area in which the individual has a legitimate expectation of privacy.

Therefore, a company, with a written drug policy that defines the company's prohibitions and actions, clearly articulated to the workers, retains the right to conduct random searches on the basis of its obligation to provide a safe work site.

The Justice Department, in a recent "Friend of the Court Brief" stated that Fourth Amendment rights are not implicated in drug testing by private employers. That Justice Department position was based on the following:

1. In the workplace, legitimate expectations of privacy are limited.
2. Unlike blood testing, unobserved urine testing is considered by the courts to be neither a search nor a seizure.
3. Testing for drugs under an established program can be considered as a condition of employment. Employees, therefore, necessarily consent to testing by their conduct when they agree to work under a known and established drug program.

Due Process Provisions

Challenges to drug testing have also been raised on the issue of "due process," under the Fifth and Fourteenth Amendments of the Constitution. These amendments refer to the government's requirement to provide due process before a person is deprived of life, liberty, or property.

Due process claims in drug-test situations are usually predicated on the supposed inaccuracies of the tests and the lack of proven impairment.

Numerous courts have upheld the accuracy of screen tests like EMIT, with confirmation tests of positive results by G-Mass Spectrometer. Therefore, this testing criteria is admissible as evidence and acceptable to the courts.

In any case, private employers are not strictly bound by the constitutional guarantees of due process. Employers would, however, be advised to consider what the courts have labeled as proper and reasonable when conducting testing programs.

ADDITIONAL LEGAL CONSIDERATIONS FOR DRUG TESTING

The following laws have an impact on company drug testing programs:

1. *The Federal Privacy Act of 1974.* Although this applies only to federal agencies, this act is now a model for state privacy statutes. Under this act, employees are allowed access to their records. Unwarranted and unauthorized access and disclosure to anyone not entitled is prohibited. Employees may bring civil suit for intentional or willful violations of this act.
2. *The Federal Omnibus Crime Control Act.* This act prohibits eavesdropping on oral or transmitted communications. It may have an impact on the conduct of an employer's undercover drug operations.
3. *State Common Law.* Under tort law, persons have the right to sue to collect damages created by willful or negligent civil acts.

   Under common law, an invasion of privacy may occur when:

   a. Someone appropriates or misuses another's name.
   b. A public disclosure of true private facts damages someone's reputation. Under this claim, release of positive

test results to unauthorized parties whether the information is true or false could subject the employer to suit.

c. Information places a person in a false light in the public eye. Under this claim, an incorrect drug-test result or an unfounded accusation of drug use or impairment could subject the employer to suit.
d. Intrusion occurs into another person's seclusion. This may apply in cases where drug searches are unannounced and against reasonable privacy expectations.

An employer can be held responsible for any actions against an employee that causes harm. These may include negligent claims involving hiring, testing, and wrongful discharge.

4. *Handicapped discrimination.* Under federal and some state handicap laws, alcoholism and drug addiction are protected by the "handicap discrimination" law. This law states, "an employer may not discharge or otherwise affect the employment of the alcoholic or drug addict unless the employee cannot perform his or her job, or the employee presents a safety threat to fellow employees or property." But, even when the drug or alcohol addicted employee cannot perform the job or is a safety threat, the employer must make "reasonable accommodations" for the employee. Presumably, an employer "reasonably accommodates" a drug or alcohol addict by allowing the employee rehabilitation opportunities.

   A majority of the states have such handicap discrimination legislation. The federal law applies to government agencies and contractors on federal government projects if the amount of the contract is more than $2,500.00.

Arbitration decisions accept the requirements of drug screens as legitimate exercise of management's responsibility to maintain safety, efficiency, and discipline.

The employer's right to request a drug screen is acceptable, but the circumstances related to that screen and the use of the results may be closely scrutinized.

A number of public sector cases have addressed the question of whether the drug-testing results are evidence of impairment.

Most have agreed that the accepted test methods in use today do not measure a level of impairment. Only the presence of a substance or metabolite associated with drug use is detected.

A positive drug-test result only proves that an individual has used a drug. As we mentioned previously, a company policy mandating a "zero level" of drug use avoids the pitfalls of dealing with the impairment issue.

Preemployment Testing

Preemployment drug screens do not violate most existing state or federal statutes, or any common-law doctrine if performed in a reasonable and unbiased manner. Employers have the right to require applicants to be drug free as a condition of employment.

Again, action against an applicant should be taken only when a positive screen test is confirmed by a second test method. Many employers require preemployment drug screening but do not cite a positive confirmed test result as the reason for nonhiring. They avoid the issue by simply indicating that the applicant did not meet the requirements for the position.

Negligence Claims

Negligence claims can be brought against the private employer as well as the government. There are three basic types:

1. The employer may be liable for negligence in hiring a substance abuser who harms another person while performing his or her job.
2. An employer may be liable for negligence if he or she fails to conduct a drug screening procedure with due care.
3. An employer who maliciously spreads untrue reports of positive test results could be sued for libel and slander.

These types of negligence claims can involve:

1. *Negligent hiring.* A state court decision recently awarded damages to a boy who was sexually assaulted by an intoxicated hotel employee. The employee had previously been fired from his job as a dishwasher because of drinking. The hotel later rehired him, even though they knew that he continued to drink on the job.

2. *Negligent testing.* Employers should be sure that the laboratories they hire or the technicians they use to perform such tests are well qualified and meet high-quality control standards.

   If on-site testing is done using company's employees as technicians, then the employer must ensure that these employees have been properly trained in test administration and know how to protect the chain of custody for samples.
3. *Libel and slander.* An employer who spreads information about positive test results that later prove to be false may be found to have committed libel or slander, even if his or her actions were accidental or merely careless.

   Employers are responsible to ensure that information on drug-testing results are kept confidential. Access to test results should strictly be maintained on a "need-to-know" basis.

THE OMNIBUS DRUG ACT OF 1988

As part of the Omnibus Drug Initiative of 1988, Congress passed the Drug-Free Work Place Act. Under that Act, federal contracts or grants over $25,000 will contain mandatory provisions for the maintenance of a drug-free workplace.

Contractors or grantees who fail to set up or maintain those provisions may face suspension of payments, termination of contract, or debarment.

The Act requires a written employer drug policy and an employee educational program. The Act does not address the issue of workplace drug testing.

The Act's Main Provisions

Federal contracts or grants after March 18, 1989 may contain a contract clause that will require of the contractor or grantee:

1. Certification of the maintenance of a drug-free workplace as part of the contract or grant.
2. Development of a drug-free workplace statement, and publication and distribution of a copy to each employee. That statement will notify employees about workplace

prohibitions on the possession, manufacture, distribution, or sale of illegal drugs. It will also contain information on the company's employee drug-information program and any disciplinary actions for violations of the policy. Employees must, as a condition of employment, abide by that statement and inform their employer of any workplace drug convictions.

3. The institution of an Employee Drug Information Program. This program will inform employees about the dangers of workplace drug use. It will include information on the company drug policy and any penalties for violations. This program must also inform employees about any available drug treatment, counseling, or assistance programs.
4. Establishment of a disciplinary program for violations. Employees convicted of a workplace drug offense have two choices. They can participate in a drug assistance or rehabilitation program, or face sanctions by the employer. These sanctions, up to and including discharge, or the voluntary participation in a treatment program, must occur within 30 days after the employee tells his or her employer about a conviction for a workplace drug offense. A sample employee disciplinary program is outlined in Table 10–8.
5. Notification of workplace drug convictions. Workers must inform their employers within five days of a conviction for violation of a criminal drug statute occurring in the workplace. Employers then have ten days to notify the federal contract or grant agency.
6. Adherence to good-faith effort. Contractors or grantees must continue a good-faith effort to maintain a drug-free workplace.

A revision of the Federal Acquisition Regulations is underway and will include rules for the conduct of suspension and debarment proceedings against contractors.

Any proceedings against employers for violations of the Drug-Free Work Place Act must be initiated by the contract/grant officer and forwarded to the agency head in writing.

Proceedings against employers are based on three possible causes:

Table 10–8 Sample Company Disciplinary Program

1. An applicant with a positive preemployment confirmed drug test is denied employment for a minimum of 30 days.
2. A positive drug-test result (first offense) results in 30 days suspension without pay and mandatory participation in follow-up testing program.
3. Follow-up testing is assigned by lottery at the employer's discretion. Any positive test results mean immediate discharge.
4. A positive drug-test result (second offense) results in immediate discharge.
5. Employees under an initial amnesty period or when voluntarily seeking assistance for a substance-abuse problem are referred to a company Employee Assistance Program (EAP) for rehabilitation and counseling.

Companies may wish to modify the disciplinary program to fit their particular needs. Consideration should be given to established grievance procedures and existing collective-bargaining agreements. All employees should receive copies of the company's drug policy, including the disciplinary guidelines established.

---

1. The contractor or individual receiving a grant has made a false certification that they will maintain a drug-free workplace.
2. Violations of any of the requirements outlined in The Drug-Free Work Place Act of 1988.
3. If "such a number" of employees is convicted of violations of criminal drug statutes occurring in the workplace a good-faith effort is not being maintained.

No suspension, termination, or debarment actions against a contractor or grantee will occur until after a hearing and the issuing of final determination. Employers have the right to judicial review of that determination.

The head of a federal agency has sole authority to waive any actions against an employer for violations. This waiver can occur whenever the agency head feels it would be in the interests of the agency, federal government, or public.

Drug-Testing Programs and The Drug-Free Work Place Act

Congress did not address the issue of drug testing within the Drug-Free Work Place Act. In fact, the Act neither requires nor prohibits employee drug testing.

This lack of drug-testing criteria is a serious flaw to the Act's effectiveness. Congress sidestepped the controversy surrounding drug testing and put the testing problem back on the employers.

Employers, therefore, find themselves in a "Catch-22" situation. They are responsible for maintaining a drug-free work place; yet, they are not given the right to use drug testing to identify or control workplace drug use.

Each federal agency is free to impose additional requirements on contractors or grantees, including drug testing. This ensures that the implementation of the Drug-Free Work Place Act will differ from one federal agency to another.

Additional Agency Requirements

The Drug-Free Work Place Act does not prevent any federal agency from enacting a more stringent drug-control policy. Some federal agencies are, therefore, incorporating established federal drug-testing guidelines in their contract/grant requirements.

Federal guidelines allow drug testing on preemployment, postaccident, for cause, and on a random basis. Testing is allowed whenever there is reason to suspect drug use. Also, employees engaged in work affecting national security or the safety of the workplace or public can be tested.

Pass Along and Trickle-Down Effect

The requirements of the Drug-Free Work Place Act of 1988 apply only to direct participants in federal grants or contracts. They do not apply to federal-assisted work. It is, however, likely that a pass-along or trickle-down effect will occur.

Responsible parties in federal grants or contracts may start to require that all employers involved in a project adhere to the Drug-Free Work Place Act requirements. For example, a general contractor or construction manager required under federal law to maintain a drug-free workplace is likely to require other employers and subcontractors on that project to abide by the Act.

This pass-along requirement will have a larger impact on the number of employers affected by the Act than the limited coverage the Act itself implies. Bureaucratic overcautiousness of federal agencies and employer general business practices will also increase the Act's impact.

Some federal agencies are already applying the Drug-Free Work Place requirements with a broader interpretation. They are requiring all their suppliers of goods or services to comply with the Drug-Free Work Place Act, even for purchases from vendors that may exceed the $25,000 limit over the course of a year.

Employers also are setting up a company-wide drug-free workplace program instead of a separate program for each federal and nonfederal contract.

Impact on The National Labor Relations Act (NLRB)

The language of the Drug-Free Work Place Act and the congressional debate surrounding its passage shows that the intent of congress was not to circumvent the NLRB.

It is unlikely that an employee bargaining group could prevent the employer from complying with the Act's requirements. Such action would likely be considered a violation of the NLRB requirement to bargain in good faith. Some unions have, in fact, already issued statements permitting drug testing when required by owners or government regulations.

The NLRB normally requires employers to bargain with their unions over terms and conditions of employment. Employers with collective bargaining agreements should include compliance with the Act within their contract negotiations. This is particularly true with reference to the type and method of employee disciplinary actions proposed for workplace drug offenses.

Costs Associated with Employee Assistance Programs

Employers must inform employees of any available drug assistance, rehabilitation, or counseling programs. The language of the Act does not mandate any employer financial responsibility for such programs.

Employers may simply make information available to employees on local drug-treatment programs. Providing drug treatment program brochures and contact information on site is one method to comply with this requirement.

Compliance Assistance

Employers can obtain compliance help from a number of federal, state, and local sources.

Federal agencies have established contact personnel for information on the Drug-Free Work Place Act. Call or write to any agency's Drug-Free Work Place Act Coordinator to answer questions about the program instituted at that particular federal agency. The February 1989 Federal Register published a list of federal agency DFWA contact personnel.

The National Institute on Drug Abuse (NIDA) has established two toll-free phone lines to provide information and compliance assistance. They are the Drug-Free Work Place Helpline (1-800-843-4971) and the Drug Abuse Information and Treatment Referral Hotline (1-800-662-HELP). These services are available during normal business hours (8:30 A.M. to 4:30 P.M., Eastern Standard Time). NIDA also publishes a series of information sheets, booklets, and reports on drugs and drug counseling topics. These materials are useful as a reference, or as employee instructional materials.

Private consultants, hospitals, state, and local health or drug-abuse agencies provide a valuable resource for information. They can be of assistance with your company's employee training and information sessions. Check the listings in your local phone directories.

Information, programs, and publications are also available through the National Safety Council and some industry trade associations.

Impact of the Drug-Free Work Place Act

The Act represents congressional acknowledgment of the existence of a workplace drug abuse problem in this country.

As a first step, it helps to focus attention on that issue. It is not, however, a solution. More positive action to identify drug abusers and to control drug use is needed.

The lack of the Act's drug-testing requirements hampers the employer's attempt to control on-the-job drug abuse through testing programs. Drug testing identifies drug users and helps determine the extent of workplace drug use.

Identifying a drug abuser is a critical step in providing rehabilitation and treatment—before it is too late. Education and information on drug use is important, but it cannot force an abuser to seek treatment. Nor can it make a drug abuser break the addiction cycle. Anyone in the workplace who is currently using illegal drugs is unlikely to be persuaded to stop those actions based on the requirements of this legislation.

For employers, it represents the imposition of additional paperwork and training burdens, without any measured return for their investment. For employees, it only ensures the receipt of drug abuse and treatment information already widely available. For taxpayers, it represents an increase in the cost of industry doing business with the federal government.

The extent to which this legislation will reduce drug use remains to be seen. However, its success will be difficult to measure. We currently do not have an accurate baseline to access the amount of workplace drug abuse that occurs or what that abuse actually costs this country each year.

The amount of employers and employees sanctioned under the Drug-Free Work Place Act will be limited. Those sanctions are triggered in part by convictions for violations of criminal drug statutes occurring in the workplace. On-the-job site convictions are rare, and convictions that occur off the job do not count under the Act's provisions.

EXECUTIVE ORDER 12565 ON DRUG-FREE FEDERAL WORKPLACES

This order was signed by former President Ronald Reagan, and became effective on September 15, 1986. The President deemed it necessary to issue this executive order based on the findings from the "White House Report on a Conference for a Drug Free America."

Under this order, federal employees are required to refrain from using illegal drugs, on or off duty.

Each federal agency must develop a plan for achieving a drug-free workplace through:

1. Employee assistance programs.
2. Supervisory training to assist in identifying and addressing illegal drug use.
3. Programs involving self-referrals as well as supervisory referrals to treatment.
4. The identification of illegal drug users through testing on a controlled and carefully monitored basis.

Drug testing may be conducted where:

1. There is a reasonable suspicion that an employee uses illegal drugs.
2. There is an investigation regarding an accident or unsafe practice.
3. There is counseling for illegal drug use through an Employee Assistant Program (EAP).
4. There is a policy to test any applicant before employment.

Each federal agency must also establish a drug program to test for the use of illegal drugs by employees in "sensitive positions" or positions affecting workplace safety and health.

All drug-testing programs must protect the confidentiality of test results and related records. The Secretary of Health and Human Services has issued guidelines for federal drug-testing programs. All federal agencies will conduct such testing according to these guidelines.

Agencies may discharge any employee who uses illegal drugs, refuses to obtain counseling or rehabilitation, or continues to use illegal drugs. However, an agency is not required to discipline any employee who voluntarily admits to being a user of illegal drugs, obtains counseling or rehabilitation, and refrains from using illegal drugs.

The importance of Executive Order 12565 is that it is a federal government recognition of the existence of drug problems in the workplace. It further recognizes the need for testing programs to control that drug problem. The Executive Order provides guidelines that assist private employers in developing their own company drug policy and testing programs.

Chapter 11

Occupational Health Hazards

INTRODUCTION

Chapter 10 discusses some occupational health concerns that relate to construction work. At first glance, hepatitis may seem an unlikely construction problem and Acquired Immune Deficiency Syndrome, or AIDS, something totally unrelated. Ask yourself, does my work operation ever involve sewer, septic, or waste pipe work? Is there a chance that an injury on my site could result in someone being exposed to blood or other body fluids? If so you have a potential exposure problem.

If you use or store chemicals on your work site, spill or accidental releases are a possible hazard. This chapter also discusses techniques for chemical-spill control and clean up. With the growing awareness of chemical hazards and occupational health problems, these are areas with which construction companies need to become familiar. This chapter provides some basic information for your consideration.

HEPATITIS—BACKGROUND INFORMATION

Hepatitis, meaning "inflammation of the liver," is a condition caused by drugs, toxins, autoimmune diseases, and infectious agents.

The most common cause of hepatitis is a virus. Four types of viral hepatitis are found in the United States:

- Hepatitis A, also known as viral hepatitis.
- Hepatitis B, also called serum hepatitis.
- Delta hepatitis.
- Non-A/Non-B hepatitis.

Hepatitis A is spread by fecal (human waste) contamination or contact. Infection can occur by a number of direct and indirect routes, including oral-fecal contact, water contamination, and shellfish contamination. The incubation period for hepatitis A is three to six weeks.

Hepatitis B is caused by the hepatitis B virus (HBV), also called the Dane particle. Carried in blood, blood products, and body fluids, this virus replicates in the host's liver cells. The incubation period is three to six months.

HBV infections are spread by many routes; for example:

- Direct inoculation through the skin with an infected needle.
- Contact with infected material and the mucous membranes.
- Contact with infected blood and damaged skin.
- Sexual transmission.
- Transmission from mother to fetus.

Blood and blood-derived body fluids from an infected person contain the highest quantities of the virus. One millimeter of infected blood may contain up to 100 million infectious doses of HBV.

Blood transfusions, needle sharing by IV drug users, and needle stick injuries in the health-care field are the most common transmission routes of the virus.

Certain other body fluids, like saliva and semen, can contain the virus, but at much lower levels. (Estimates indicate a 1,000-fold decrease in viral concentrations in these materials.) Urine and feces usually contains only small concentrations of the HBV virus unless the sample is visibly contaminated with infected blood.

Delta Hepatitis affects persons already infected with HBV and can increase the severity of liver damage in HVB patients.

Non-A/Non-B hepatitis is a name given to a group of diseases caused by viral agents that exhibit hepatitis symptoms but cannot be identified as other forms of hepatitis infections. Serological tests for this disease are not currently available.[1]

TESTING FOR HEPATITIS

Laboratory Testing

Currently available testing for hepatitis B infection involves examination of a blood sample for the presence of a lipoprotein called hepatitis B–surface antigen (HBsAg), also called the Australian antigen.

This HBsAg is produced in great quantities in the liver cells of an individual infected with hepatitis B. The presence of HBsAg in the blood indicates that an individual is currently infected with hepatitis B and is potentially infectious to others.

The American Red Cross recommends that as a protective measure, blood samples from all potential victims be routinely taken as soon as possible after suspected hepatitis contamination occurrences. The samples can be preserved for later testing or for any employee-requested testing. Samples should be maintained for a minimum five-year period.

The risk for work-related hepatitis infection is well documented in the health-care field. OSHA, however, states that the risk of hepatitis B infection is not confined to just one industry group but is present whenever infected blood or other infected materials are present in the workplace.

Laboratory Test Costs

The Centers for Disease Control in 1989 cited an average per-unit cost for the laboratory HBV antibody test of $24.50 per test. Tests are conducted on blood samples when the victim exhibits symptoms of infection or when populations are considered at risk. Inoculations can be given to potentially exposed workers, but the costs are high. OSHA cites a per unit cost for administering HBV immune globulin inoculations of $211.00. Prices may vary in your area.

[1] Laboratory tests to identify Non-A/Non-B hepatitis are under development with some progress recently reported in medical journals.

Field Testing

Concern has been expressed by workers involved in sewer and waste-water repair work about the possibility of occupational exposures to infectious diseases like hepatitis A and B.

Although there are no known documented cases of infection via the exposure of workers to infected waste materials, the possibility of infection does exist.

Hepatitis A is carried in fecal (human waste) matter. Therefore, any untreated waste may be a potential source for contamination. Transmission of the virus from the contaminated source to the individual is by an oral route. The person may drink or eat from a contaminated source such as water or shellfish. A splash of contaminated waste into the face and mouth, or contact with the mouth and contaminated hands or other contaminated surfaces is another potential source of worker infection.

Hepatitis B is a blood-borne infection. Blood and related body fluids carry the virus. Hepatitis B does not live long in a water environment, but water and waste lines can be contaminated if blood and body fluids are introduced into the water or waste system.

Contact of the contaminated material with the eyes, mouth, or broken and damaged skin are possible infection sources.

Field tests can detect the presence of hepatitis viruses in waste and waste water, but these tests may not assist in documenting the cause of an individual's infection.

As stated earlier, the hepatitis A virus incubation period is three to six weeks, and the hepatitis B virus incubation period is three to six months. This lag time between infection and symptoms or diagnosis renders after-the-fact tests unreliable in determining the source of the infection.

Tests conducted after the infection is diagnosed cannot reflect the conditions existing at the time of the worker's infectious contact.

A positive test (the virus is present) taken after the fact does not prove that this contamination source existed at the time of initial infection or is the one responsible for the infection.

A negative test (no virus present) taken after the fact does not prove that the virus was not there when the infection occurred. Only a test taken at or close to the time the infectious contact occurs can pinpoint the infection source.

Tests of untreated waste-water samples are likely to be positive for the hepatitis A virus. However, a positive test on a common collection pipe does not identify the one source where the infection originated. It is difficult, therefore, to pinpoint the party responsibility for a worker's infection. The employer in most cases will end up paying the bills for exposed workers.

METHODS OF CONTROL

Hepatitis A patients are initially hospitalized on an emergency basis for support therapy and treatment. Risks of hemorrhage exist in infected persons due to liver damage from the virus.

Early intervention in suspected hepatitis B-exposed patients involves the use of one of two currently available vaccines. A treatment regime of injections is given in three doses over a six-month period. These vaccines induce protective antibody levels in 85 percent to 97 percent of healthy adults. Protection by immunization is considered effective for at least a seven-year period after treatment.

Treatment may involve hospitalization on a nonemergency basis for testing to determine the extent of liver damage and types of support therapy to be used.

OSHA recommends that persons planning HB vaccine programs consider the need for prevaccination testing for the HVB antibody. This testing identifies individuals previously infected who, because of an immunity, do not need the vaccine. Depending on the cost of inoculation in your area, this procedure could be cost effective.

The Centers for Disease Control (CDC)

The Centers for Disease Control recommends the following precautions in dealing with potential infection contaminant sources:

1. Treat all waste and blood exposures as potential contamination sources. Use protective equipment.
2. Use rubber gloves and surgical masks in accident emergency treatment where blood and body fluids are present.
3. Be sure to dispose of gloves, masks, and other infected materials properly. Seal them in plastic bags, labeled for bio-waste, and follow all federal, state, and local regulations for disposal.

4. Report and record all potential sources of infectious contact. Include the names of all persons involved, dates, and times.
5. Provide all employees with training on the proper protective equipment use, safe work procedures, and potential hazards.

Construction Trade Associations

A major construction trade association provided the following recommendations to workers engaged in waste and waste-water repair work.

1. Treat all exposure to waste water and waste-pipe materials as potential infection sources. Whenever possible, flush and drain pipes before work starts.
2. Wear rain gear, face shields, boots, and rubber gloves when working on waste pipes.
3. Discard or disinfect all personal protective equipment tools and clothing (with a 1-to-10 solution of household bleach in water) after use.
4. Practice good personal hygiene. Wash your hands and face with soap and water after contact with waste water and waste materials. Consider the use of shower facilities at waste-water work sites.
5. Rinse the eyes under running tepid water for a minimum of 15 minutes whenever waste water or waste material splashes in the eyes.
6. Report and treat all injuries that occur while engaged in waste-water work.
7. Provide training to workers on safe work practices and the health hazards associated with this type of work.

ACQUIRED IMMUNE DEFICIENCY SYNDROME (AIDS)

AIDS is a blood-borne disease caused by a virus. Infection with the virus suppresses the normal functions of the human immune system, leaving the infected person subject to a number of debilitating and fatal diseases.

According to the Centers for Disease Control, there are four major routes of AIDS infection transmission:

1. Sexual contact with an infected person.
2. Penetration through the skin by an object or substance contaminated with the AIDS virus.
3. The transfusion of infected blood.
4. Transmission from the mother to the fetus during pregnancy.

In the construction workplace, contact with infected blood or infected materials is the likeliest method of AIDS transmission.

Contact with infected blood or body fluids might occur in any industrial accident situation, presenting a danger to personnel assisting or treating injured workers.

Waste-material contact might occur in renovation or repair work in facilities where such waste are generated, like hospitals, clinics, bloodbanks, or medical-waste disposal facilities. Potential exposures also exist in repair work on existing sewer and waste-water lines.

With the growing awareness and concern for blood-born infections, companies should establish written procedures for safely treating accident victims while preventing infection. These procedures or guidelines can utilize the same safety procedures established by OSHA for Blood-Borne Infectious Diseases control outlined below.

Policies should also be established for protective equipment use and safe work procedures for all waste and sewer pipe repair work.

SAMPLE WORKPLACE PROTECTIVE MEASURES FOR INFECTIOUS DISEASES CONTROL

1. Provide educational programs and information to workers potentially exposed to infection.
2. Wear rubber gloves when handling blood, body fluids, and waste materials or treating accident victims. Surgical masks provide additional protection when working on accident victims. Dispose of used protective equipment and medical wastes properly.
3. Wear eye protection, rain gear, and filter-type masks when working on sewer and waste-water lines.
4. To disinfect gear or environmental surfaces suspected of contamination, use a bleach solution of approved bacteri-

cide or germicide. A prepared solution of household bleach & water (one part) bleach to (ten parts) water can be used as a disinfectant.

5. Use alternative methods for mouth-to-mouth resuscitation whenever possible. These might include Ambu Bags, Power Respirators, as well as a variety of resuscitation tubes or shields now being marketed. These tubes and shields provide a sanitized barrier between the person performing resuscitation and the victim.

AIDS AND COMPANY PERSONNEL POLICIES

AIDS infection is a protected handicap under the Handicap Discrimination and Fair Labor Laws. As such, an AIDS-infected person cannot be denied employment, terminated, or discriminated against solely because of the affliction.

Perhaps, the biggest barrier to overcome in working with an AIDS-infected person is other workers' fears. Casual contact has not been shown to be a method for transmission of this disease. People's unfounded fears and lack of factual information about AIDS create this unwarranted fear reaction. A company education and information program dispels prejudices and misinformed fears about AIDS.

Your company's personnel policies may need to be rewritten to address the AIDS issue. Guidelines for an AIDS personnel policy are available from the American Occupational Medical Association and from the Centers for Disease Control.

HAZARDOUS SPILL CONTROL

Whenever hazardous chemicals are used, stored, or transported, a potential exists for a spill and exposure of workers and the public to harmful chemicals. The use of chemicals also raises many environmental concerns. The dangerous effects of chemical spills on people and the environment and their liabilities can be reduced by the introduction of a company hazardous spill-control plan.

The control of hazardous chemical spills involves several distinct phases:

- Control of the hazardous material.
- Containment of the spill.
- Removal or mitigation.
- Disposal.

Hazardous materials occur in gaseous, liquid, and solid states. Each state presents a specific set of problems in dealing with spill control.

The state of the material spilled and its physical characteristics, coupled with the environmental circumstances of the spill, will determine the clean-up actions necessary. For example, hazardous material spilled into water as liquids or solids can be further classified as soluble (dissolving) and insoluble (nondissolving).

Insoluble materials can further be classified as sinkers or floaters. This, of course, refers to their ability to float on the water or sink to the bottom.

The initial and principal concern in hazardous chemical spills is the control of the immediate threat of exposure to workers and the public within the spill zone.

Depending on the classification of the spilled materials, the degree of harm possible, and the circumstances of the spill, a limited or complete evacuation of the spill area may be necessary.

All response efforts must be conducted using proper personal protective equipment with respect to the materials and hazards involved. Hazard response managers and personnel will need to consider the following needs:

- Respiratory protection.
- Chemical resistant clothing, including suits, boots, and gloves.
- Monitoring equipment.
- Spill containment and clean-up supplies.

Containing Spills

The first phase in spill-control operations is to contain the initial spill as much as possible. Each type of spill presents it own unique containment circumstances.

Gases

Gases, liquids, or solids which volitize (become vapors) are extremely difficult to contain in a spill or release situation. In most cases, the practical approach is to allow the material to safely disperse into the atmosphere.

Where gases are highly toxic (such as hydrogen sulfide or chlorine), attempts to reduce the rate of dispersion are used. This is generally done through the application of firefighting foams to the spill area. The foam reduces the amount of spilled material exposed to air and therefore the rate of dispersion.

There are many type of firefighting foams available that can be used:

1. Protein foams are made from hydrolyzed proteins and additives to prevent breakdown.
2. Floro-protein foams are protein-based foams with fluorine compounds added to make them resistant to chemical breakdown.
3. Alcohol-based foams are made to be used against polar compounds.
4. Aqueous film-forming foams (AFFFs) are fluorosurfacants (compounds with very low surface tension). These surfactant, detergent foams have the ability to expand rapidly (up to 500 times in volume) to cover an effected area.

To be effective, all foams must have a low water loss (drainage rate) and a slow collapse (breakdown rate). Foam should only be used on spill materials whose vapors are lighter than air. With heavier-than-air materials, vapors will become trapped in low-lying areas and run out from the end of the foam barrier.

Foams used on water-reactive materials, such as chlorine and ammonia, result in a short-term exaggeration of the vapor hazard, but a long-term reduction of vapor production.

Solids on Land

A solid material spilled on land is generally immobile. In some cases, lightweight, small particle-sized materials may be dispersed by the wind.

These kinds of materials can be contained by covering them with a plastic sheet. If the material does not react with water, they can be dampened with water-hose spray to increase their weight.

Materials that sublimate (change to gases) must be treated initially as gaseous spills if their vapors are hazardous.

Solids, when contained, can be neutralized or recaptured and placed into properly labeled and sealed containers.

Liquids on Land

The containment of a liquid spill on land is difficult because liquids flow along the ground slope and penetrate or are absorbed into the ground vertically.

The spreading of a liquid spill is contained by diking (building a wall to contain the material) or sumping (digging a hole lined with impervious materials in which to collect the spill).

Penetration of materials into the ground involving potential ground water-system contamination will be discussed later in this chapter.

Contaminated soils in general must be treated as solid spills and removed for containment or disposal.

A number of commercial products are available for diking liquid spills, including:

1. A two-part urethane foam material that can be used to build an impervious containment dike. This material is mixed and placed into position; it then hardens to form a barrier. This foam material can be used to plug pipes, but its ability to adhere to wet surfaces is poor.
2. A line of inflatable bags in various sizes are available to create dikes or seal off pipes. The bag is placed or inserted and inflated to form a wall or plug.
3. Another commonly used diking product is a mixture of bentonite clay and starch polymers. The polymers have a great water-absorbing capacity. This material is spread in its dry form ahead of the spill flow and sprayed with water. It then absorbs the water to form an impervious barrier.

An important point in the containment of liquid spills on land is the sealing off of all areas where the spill could enter water, sewage, or drainage systems.

Drain openings can be covered by sheets of plastic weighted

down by sand or other materials to divert the spill flow. Entry points can also be plugged with the diking products listed above.

Liquid spills can also be controlled by pumping the liquid to a temporary storage area, impervious to the chemicals involved.

Spills into Surface Water

Spills of soluble material into water presents an almost impossible containment situation.

Spills into quiet waters may sometimes be controlled by a pump and filter method, where the water is extracted, filtered, and returned.

Spills into moving waters are generally controlled by diverting the water flow to a holding pond, containment area, or more isolated water area.

Insoluble materials that are heavier than water will sink to the bottom. Containment options are limited and based upon the rate at which the materials spread or are carried by water currents.

In some spill cases, underwater barriers and dams have been erected to contain small spills. An alternative method has been to divert a stream to bypass the contaminated area, allowing the dewatering and removal of the spilled materials.

Insoluble materials that are lighter that water will float. Containment and recovery will depend on the ability to restrict initial water current dispersal and movement.

This is usually accomplished through the use of floatation collars or booms to restrict the spilled materials movement. This is a common containment method in oil spills. Materials are then neutralized, skimmed, or pump-recovered and placed into proper containers.

Ground-Water Spills

Ground-water contamination results from spilled surface materials: penetrating the ground, entering drainage and sewage systems, or leaching from land by water run-off.

Spills of soluble material that reach ground water are dissolved and carried throughout the ground-water system. Control of such contamination is difficult, involving pumping and filtering techniques or slurry trench methods for removal.

The basic idea in these situations is to eliminate the contamination source by intercepting or isolating the contaminated water. Either channeling or filter methods are used to remove contaminate materials from the ground-water system.

Removal or Mitigation

The second phase of spill control involves removal of the spill material or neutralization of the spilled material by a number of methods, including:

1. *Dilution.* The most practical solution to many hazardous spills is a dilution to a level below maximum safe exposure concentration. This technique generally is used with materials that disperse readily in air or materials that are soluble in water.
2. *Removal.* Spills that have occurred on land resulting in contaminated soil may be removed mechanically by excavation. Solids and liquids that do not penetrate the ground or floor materials may be vacuumed or pumped to suitable storage containers.

   Spills of materials that float on water may be removed through skimmers or pump filters and stored in suitable containers.

   Insoluble liquids, heavier than water, that penetrate into the soil will sink until they reach an impervious layer. These types of spills are generally removed by well-drilling and pumping operations.
3. *Sorbents.* Sorbents are materials that absorb spilled liquids. Commonly these include materials such as pearlite, vermiculite, clay, volcanic ash, and talc.

   These materials are generally spread on a liquid spill where they absorb the spilled material and, in turn, can be recovered later for disposal.

   A sorbent used for a broad range of hazardous material is made of inert silicate glass granules. As a general rule of thumb, one-half cubic foot of this material will absorb approximately two gallons of liquid.

When a spill occurs in a solution with water, sorbents called *super-slurpers* are generally used. These materials, made of synthetic polymers, will generally absorb up to two hundred times their weight in water.

4. *Thermal Oxidation.* Burning is an effective method of destroying materials in a spill situation, providing no thermal-related toxic materials are produced.

   In general, when highly flammable materials or highly toxic materials are involved in a fire, it is preferable to let the spill burn itself out. Attempts to put the fire out may result in incomplete combustion and the production of additional toxic materials.

5. *Neutralization.* Chemical substances can be broadly classified as either acids or bases. Specific materials can be used to neutralize (make less harmful) many hazardous chemicals.

   Neutralization is an exact process, difficult to produce even in a well-equipped laboratory. It is not generally used in field situations.

   In the field, the neutralizing chemical must be readily available in large quantities and must be weak and not harmful to the environment or workers.

   Haphazard neutralization may produce chemically reactive products more dangerous than the initial spill.

   The easiest example of a common neutralization process is the action of baking soda on a mild acid like vinegar. This common high school chemistry neutralization produces large quantities of carbon dioxide and carbon monoxide gases. In small amounts, this procedure is safe; in large quantities and in contained spaces, this procedure could be harmful. The amount of carbon dioxide gas produced could reduce ambient oxygen levels.

6. *Biological Degradation.* Nature normally biodegrades, or breaks down, all material, but this process is slow. Research has developed numerous microbes that can be utilized in biodegrading specific spilled materials.

   This process has been used recently on some oil-tanker spills with limited success.

   Biodegradable processes are at best slow and difficult to use in field situations. For example, one EPA approved technique used to degrade gasoline requires the proper mixture of 31 different microorganisms with sufficient nutrients and oxygen. The procedure is therefore expensive and difficult to control.

7. *Encapsulation.* This provides a method of controlling a toxic spill by surrounding the hazardous material with an inert material. The most common application of this technique is the removal of asbestos from buildings. Friable asbestos is coated with an epoxy mixture before removal to prevent fibers from entering the air.

   Problems with this process involve the creation of large quantities of encapsulated materials that require long-term storage.

Disposal

Unless a spill can be recovered in a form that will permit its reuse, it is a hazardous waste that creates its own disposal problems.

In an ideal situation, these hazardous waste materials could be transformed into usable products, detoxified, or completely destroyed. Unfortunately, facilities to achieve these results are expensive to use and limited in number.

Most hazardous materials are subject to long-term storage with a host of regulatory processes and costs involved in their safe handling.

Waste-disposal techniques, including burning, neutralization, and biodegration, are used in only about two percent of all hazardous waste disposal. Long-term storage by secure land fill and deep-well injection account for most hazardous waste disposal today.

We produce hazardous waste far in excess of our ability to store or recycle the materials. The recurrent problem with illegal toxic waste dumps and the cost for clean-up is becoming a multibillion dollar problem for America.

IMPENDING LEGISLATION AND REGULATIONS IMPACTING OCCUPATIONAL HEALTH

OSHA labeled their regulatory push in the 1990s as a concentration on workplace occupational health concerns. Additional legislation will eventually expand that concern into the construction industry. These new regulatory areas include:

1. *Permissible Exposure Limits.* OSHA revised the airborne contaminant exposure levels for all the substances in the OSHA Z-Tables. They also adopted exposure limits for some 200 chemicals never regulated before.

   Although the limits at present apply to general industry only, an expansion to include construction will occur shortly.

   Although not mandated by the standard, a reduction of the permissible airborne exposure limits may force employers to conduct air-quality monitoring as a precautionary and defensive measure.
2. *Confined Spaces.* OSHA is in the process of promulgating a confined-space standard that will mandate air space testing before working in a confined space. Additional requirements will involve permit systems for confined space entry and increased educational, supervisional, equipment, and medical-monitoring requirements.
3. *Ergonomics.* The interface of man in the work environment and the design of tasks and tools to reduce occupational injuries and illness will likely be a specific concern of OSHA in the 1990.
4. *Cumulative Trauma.* A general recognition of occupational disease related to long-term work conditions is growing. This will alter the current perception of workplace injuries and illnesses as associated with a specific event or single exposure. OSHA is expected to focus on the Cumulative Trauma issue.
5. *Manmade Metallic Fibers.* A controversy over the health effects of manmade materials like fiberglass may have a profound effect on construction. Some studies maintain that all fibrous materials like asbestos can create lung scarring and cancer. The debate is still raging, but OSHA has expressed interest in this issue.
6. *Bureau of Labor Standards Accident Record Keeping.* BLS is revamping the methods they will require employers to use in recording workplace accidents and illnesses. Preliminary work by organizations like the Keystone

Center promise a much more complex record-keeping system and a broadening of what is classified as work-related injuries. More emphasis is being placed on the recording of occupationally related diseases. The BLS revision is expected to be completed in 1991.

7. *Public Protection and Disaster Planning.* The construction industry can expect to hear increased concern over the issues of protecting the public from hazards associated with construction activities and the need for disaster planning for construction projects.

   Some states, like New York, have passed laws in the wake of some tragic public construction-related accidents, mandating increase public protection.

   The American National Standards Institute A-10.33 is committed to developing a new consensus standard on public protection.

   Some or all of these standards may deal with public protection from work-generated and work-related air and water contaminations. The Community Right to Know provisions under SARA, Superfund Authorization and Redevelopment Act, already allow public access to your company's Hazardous Chemical Inventory List. Public knowledge of these hazardous chemicals may involve some employers in lawsuits for alleged damages.

Chapter 12

The Sick Building Syndrome

SICK BUILDING SYNDROME AND BUILDING-RELATED ILLNESS

Introduction

Growing awareness of the potential health problems of indoor air pollution have focused attention on the Sick Building Syndrome. More and more, contractors are being named in the lawsuits arising out of building-related health complaints.

SBS-related problems are also a potential source of work-related health problems for workers during interior finishing operations. As the construction work progresses to the enclosed structure and materials like glues, panel, finishes, etc., are used inside the structure, harmful levels of air contaminants can occur. Contractors have therefore both an immediate and long-term concern with building-related air-quality problems.

Some buildings seem to have a high number of occupant complaints, ranging from minor ailments to mysterious maladies like Legionnaire's Disease.

The National Institute of Occupational Safety and Health (NIOSH) and the Environmental Protection Agency (EPA) have estimated that 20 to 35 percent of office workers are adversely affected by the air quality at their place of work. A study by a British consultant on building use found that 80 percent of workers each year suffer from at least one symptom associated with work in a sick building environment. Other studies in Holland and Sweden point to similar findings on employee-related building complaints.

So just what is the "Sick Building Syndrome" (SBS), and what makes a building sick ? When we talk about a sick building, we are really using a generic term for any ailment related to poor Indoor Air Quality (IAQ).

The past decade's concentration on conserving energy and new construction techniques have given rise to a multitude of high-energy, efficient airtight buildings, buildings that rely upon mechanical air circulation and internally maintained temperature and humidity controls.

Offices lacking proper air circulation can generate conditions where germs, allergens, and air pollutants can be released, maintained, and circulated by the building's own air-conditioning and air-handling systems.

The collection of indoor air contaminants and the growth of organisms on environmental surfaces can lead to long-term occupant exposures to pollutants, infections, and allergens. Occupants begin to complain of a number of maladies including headaches, sore throats, flu-like symptoms, fatigue, irritation, and coughing.

When particular medical conditions like asthma, pneumonia, and respiratory infections occur, they are classed as Building-Related Illnesses (BRI).

Poor air quality is generally the result of improper circulation and the presence of harmful air-borne substances or pollutants. Air pollutants are classified into three very broad groups: particulates, biogens, and gas-like compounds.

Particulates

Particulates include small solid particles suspended in air, such as the following:

1. *Dusts.* Dust are any solid material that may be dispersed in air when the particle size is small enough; for example, metals, woods, suspended solids, dirt, and other materials may irritate the respiratory system or create allergic reactions in sensitive individuals.
2. *Insect and Animal Materials.* Insect and animal parts and wastes can be circulated by air-handling systems in a

building. These solids may irritate the eyes, nose, throat, and lungs and create sensitivity and allergic reactions.

3. *Plant Materials.* Pollens, spoors, oils, and other plant products can contaminate an air-supply system. Some, like ragweed, affect people with a specific allergy sensitivity.

Biogens

Biogens include all types of bacteria, virus, molds, biological toxins, and algaes that breed on indoor surfaces and circulate in the air, including:

1. *Legionella.* This is the infectious organism responsible for Legionnaire's Disease. Legionella can grow wherever the right temperature, humidity, and food supply exist. Legionnaire's Disease generally occurs as a type of pneumonia.
2. *Acute Respiratory Diseases.* These include such diseases as measles, varicella, humidifer disease, and Q fever. Airborne diseases can be transmitted through the air-handling systems of a building.
3. *Biological Toxins.* Many microorganisms either contain or produce toxins. As these organisms grow on many types of environmental surfaces when conditions are right, they release these toxins into the air.
4. *Biological Aerosols.* Many microbes release organic chemicals during their growth cycles. These chemicals can become airborne and circulate in a building causing irritation and discomfort. The moldy smell often mentioned in occupant complaints of sick buildings is associated with microbial contaminations.
5. *Allergens.* Exposure to a wide range of suspended solids, pollutins, and biogens, like bacteria, virus, algae, and molds, can create allergic reactions in sensitized individuals. Effects can range from the annoying to the life-threatening. Some allergic individuals may experience respiratory, eye, and nose irritation or muscle and joint pain. Others may develop specific conditions like allergic asthma or hypersensitivity pneumonia.

Gas-Like Compounds

These types of airborne contaminants number in the hundreds, from the common to the exotic, including:

1. *Formaldehyde.* This is emitted into the air from certain types of carpeting, carpet pads, paneling products, insulations, glues, pressed woods, and fabrics. Formaldehyde is an irritating compound when inhaled or in contact with skin and eyes.
2. *Radon.* This is an invisible gas produced by the decay of radium, a naturally occurring radioactive material in the earth. Radon seeps into building basements from the surrounding soil and can enter into a building's air-circulating system.
3. *Tobacco Smoke.* Second-hand smoke contains nicotine, tar, particulates, and other harmful and irritating compounds. Second-hand smoke is considered to be more irritating than tobacco smoke inhaled by the smoker. Secondary smoke can cause eye and respiratory-tract irritation.
4. *Carbon Dioxide.* Carbon dioxide is formed whenever carbon-containing products like gasoline are burned. It is also emitted in natural respiration. It is an odorless, colorless gas. Short-term exposure to carbon dioxide has been linked to headaches, dizziness, breathing difficulties, and drowsiness.
5. *Carbon Monoxide.* This gas is also produced when products containing carbon are burned or broken down. Carbon monoxide is another colorless, odorless gas. In the body, it replaces the oxygen circulating in the body, leading to death by asphyxiation.
6. *Nitrogen Dioxide.* This substance is generally produced in high-temperature combustion operations. It is a respiratory irritant.
7. *Sulfur Dioxide.* When sulfur-bearing compounds like fuel oils are burned, sulfur dioxide is released. A respiratory irritant, sulfur compounds have been linked to aggravation of preexisting lung conditions and occupationally related lung diseases.

8. *Ozone.* Ozone is created when electrical arcs pass through air. Ozone can be produced by the operation of electrical switches and circuits, as well as by fires and lightning. Ozone is another respiratory irritant.
9. *Volatile Organic Compounds.* All kinds of organic vapors are released into the air. These include vapors from chemicals used in the manufacture of various products and materials, like solvents, cleaners, and many other airborne gases, mists, fumes, and vapors.
10. *Fibers.* Friable fiber material either natural or manmade can become airborne. Probably the best known of these is asbestos. Asbestos fibers entering into the lungs create scar tissue leading to lung deficiencies and disease. Controversy exists with manmade fiber materials like fiberglass. Some scientists contend that these fibers can also damage the lungs.

    Airborne pollutants can also migrate from outside of a building and become trapped in tightly sealed buildings or in areas with poor air circulation or air exhaust characteristics.

WHEN IS A BUILDING CLASSIFIED AS SICK?

Experts classify a building as suffering from Sick Building Syndrome when the following conditions are present.

1. More than one-fifth of the building's occupants complain of SBS symptoms, including headaches, sore throats, breathing difficulties, fatigue, eye irritation, or outbreaks of flu-like symptoms.
2. The symptoms indicated above continue for a two-week or greater period.
3. The cause of such complaints or illnesses is not determined to be from another cause, like a flu epidemic, or a cause directly linked to the building (building-related illness).
4. The symptoms disappear when the occupants are outside of the building for any length of time.
5. Other environmental factors are not shown to be the cause of the problem. For example, lighting glare from reflective

or glossy painted surfaces can create eye irritation and headaches.

6. Air testing and a physical examination of the affected building confirm the presence of a contaminate capable of causing the symptoms noted. Air tests indicate the existence of poor indoor air quality.

Dismissing employees who complain of building-related health problem as complainers or hypochondriacs is unadvisable. NIOSH and other researchers have confirmed that the Sick Building Syndrome is real and that it does create illnesses in people.

Litigations related to sick buildings and building-related diseases are growing and the courts are awarding damages. Under personal injury, litigation damages can include pain, suffering, loss of capacity and consortium, mental anguish, medical costs, and loss of earnings.

A recent Worker's Compensation ruling issued by the Florida Department of Labor and Employment points out the growing awareness of SBS problems and creates a legal precedent. The University of Florida was required to pay for past and future medical bills for an illness attributed to an employee breathing contaminated building air. In addition, the cause of extremely high concentrations of molds in the building's air-handling equipment and ducts required a costly renovation of the HVAC system.

In California, a worker sued over 200 businesses because of indoor air pollution in the building where he worked. Everyone connected with the building's ownership, construction, and maintenance was named in the suit. Because of the expense involved in defending such a case, the employee walked away with over $600,000 in settlements.

WHAT CREATES THIS SICK BUILDING SYNDROME?

The National Institute of Occupational Safety and Health has investigated 529 cases of Sick Building Syndrome. NIOSH places the blame, in one-half the cases studied, on poor air ventilation.

Airborne chemicals either from outside or inside sources accounted for a quarter of the rest of the cases. Microbes accounted for 5 percent of the remaining cases and another 4 percent were labeled as undiagnosed and, therefore, mysterious.

Whereas the cause of Sick Building Syndrome and Building-Related Illness may be traced to airborne contaminants, the reasons why one building and not another develops such problems are more complex.

The Honeywell Company analyzed 30 buildings over an 18-month period. They concluded that improperly designed and maintained High-Volume Air Conditioning Systems (HVAC) are the greatest contributors to the Sick Building Syndrome (SBS).

Chemical contaminants were responsible for SBS complaints in 75 percent of the cases studied by Honeywell. Thermal problems (too hot, too cold) were cited as a contributing factor in 55 percent of the cases. Microbiological contaminants created 45 percent of SBS complaints, and humidity problems were cited in another 30 percent of the cases.

Honeywell also cited building air-condition design and maintenance problems, including the following:

Maintenance Problems Dirty air-intake systems, contaminated air ducts, clogged air filters or heating and cooling coils, and increased system inefficiency due to improper lubrication and maintenance were noted.

Design Problems

- The air-flow systems did not evolve as the building use increased. This created blocked air flow, dead air spaces, and reduced air-circulation conditions.
- Increased use of computers and copiers and other electrical equipment without providing venting to the outside was cited. This leads to air-flow systems that are overloaded and unable to provide sufficient clean air.
- Energy conservation methods added to buildings that reduced the use of outside air for ventilation. Air exchange was, therefore, inadequate.
- Some building were designed without outdoor air intakes or the air intakes allowed outside fumes like parking-lot auto-

mobile fumes to enter the building. Air intakes can also be polluted by positioning them too close to vent stacks, exhaust supplies, and outside mechanical or chemical processes.

- In others systems, inadequate amounts of outdoor air were being brought into the building. Intake vents can also be blocked by landscaping, new construction, or when original air flow systems are not upgraded to meet increased demands.
- Some buildings did not have central exhaust systems for building air. This problem was most severe when a combination of ceiling diffusers and ceiling return plenums were used. This combination, in effect, short circuited the building's air-supply system.

Kemper Research Foundation, a nonprofit biotechnology research center, studied 25 commercial buildings throughout the United States. They concluded that High-Volume Air Conditioning systems are not the root cause of most building-related illness. These air-flow systems mostly acted as air-dispersal systems to spread pollutants throughout the building.

Their research indicated that building carpets and ceiling tiles harbored large amounts of microorganisms capable of causing irritation and disease. These microbes—not organisms in the HVAC systems—were responsible for employee ills. The air-flow systems acted primarily as transport systems.

Kemper also indicates that the presence of volatile organic compounds and physical agents must be considered in assessing a building's problems.

Disinfecting contaminated carpets and ceiling tiles proved difficult. Conventional carpet cleaning and treatment with biocides proved ineffective. Conventional methods can lead to increased microbe growth and the introduction of harmful chemicals in biocide treatments.

Kemper's research lead to the recommendation to use Sylgard, a surface bonding antimicrobial agent made by Dow Chemical. The product had no toxic properties and reduced microbe populations by 80 percent to 99 percent. Sylgard was not found effective in the treatment of air ducts or air-handling equipment.

HOW IS A SICK BUILDING DIAGNOSED?

The initial investigation of a SBS complaint involves a screening process. First, consultation is conducted with the building owners and occupants in response to occupant complaints. Then a physical examination of the structure and its environment (See Figure 12–1) is conducted. At this stage, you cannot conclude the existence of a building-related problem in the absence of documented clinical evidence of health problems.

1. **ASSESSMENT SURVEY**

Client ____________________
Contract ____________________
Address ____________________
____
Phone ____________________
Date ________ 19 ________
Site ____________________
Address ____________________
Day ____________________ Time: From ________ To ________
____

2. **BUILDING EXTERIOR**

(a) Age of Structure ____________ # of Stories ________
Type of Structure ____________________
Type of Construction ____________________
____

(b) App. Dimensions: Ht. ________ Lt. ________ Wt. ________
Foundation: Yes ______ No ________ On Slab ________
Type of Slab Construction ____________________
____

(c) BUILDING ORIENTATION: Show orientation of site using an attached aerial photograph, map, or drawing. Indicate north and all relevant distances.

Figure 12–1 Physical Assessment Form (Reprinted by permission of the authors, TCR, Inc. All rights reserved.)

(d) Parking Facilities: Yes ________ No ________

Type: Exterior __________ Interior __________ Open __________

Contained __________ Above __________ Sub __________

Any Potential Sources (Describe) ______________________________

____

____

(e) Locate All Exterior Air Intakes and Exhausts:

#	**LOCATION**	**DIMENSIONS**	**TYPE**	**CONDITION**	**REF.**	**REF.**
1.						
2.						
3.						
4.						
5.						
6.						
7.						
8.						

Use additional pages as necessary.

(f) TOPOGRAPHY: Elevation ________ Prevailing Winds ________

List all Immediate Structures:

BUILDING NAME OR ADDRESS	**TYPE**	**OPERATIONS**	**SOURCE**
1.			
2.			
3.			
4.			
5.			
6.			
7.			
8.			

Use additional pages as necessary.

(g) ROOF: Type ______________________________

Any Venting: Yes __________ No __________

#	**LOCATION**	**TYPE**	**DIA.**	(LIST) **SOURCE**
1.				
2.				
3.				
4.				
5.				
6.				

Use additional pages as necessary.

Figure 12–1 ***(Continued)***

3. **HIGH VELOCITY AIR CONDITIONING**

(a) In use: Yes ________ No ________ Location ____________________

(b) Type: Make ______________________________ Model __
Plate Information ______________________________
____
____

(c) Responsibilities:
Installation ______________________________
Maintenance ______________________________
Repair ______________________________
Any Maintenance within the last six months:
Yes _____ No _____
Type ______________________________
Date ______________________________

(d) Filters: Type ______________________________
Location ______________________________
Condition ______________________________
Odor ______________________________
Observation ______________________________
____

Last Filter Change Date ____________________ How Often ___

(e) Drip Pans : Type ______________________________
Location ______________________________
Condition ______________________________
Odor ______________________________
Observations ______________________________
____

(f) Humidifier :
Type ______________________________
Make _______________ Model _______________
Location ______________________________
Condition ______________________________
Odor ______________________________
Observations ______________________________

(g) Coils, Plates, Condensers:
Type ______________________________
Make ______________________________ Model ___

Figure 12–1 (*Continued*)

Location ______________________
Condition ______________________
Odor ______________________
Observations ______________________
____

(h) Ducts: Type ______________________
Lining ______________________
Condition ______________________
Odor ______________________
Observations ______________________
____
When Installed ______________ Repaired ___
Type of Repair ______________________
By Who ______________________
Last Cleaning ______________________
Type ______________________
By Who ______________________

(i) Vents & Returns

#	LOCATION	TYPE	SIZE	CONDITION
1.				
2.				
3.				
4.				
5.				
6.				
7.				
8.				
9.				
10.				
11.				
12.				

Use Additional Pages As Necessary.

(j) Boilers & Heating Systems:

Type ______________________
Location ______________________
Condition ______________________
Odor ______________________
Observations ______________________
____

Figure 12–1 *(Continued)*

INTERIOR BUILDING LAYOUT

(a) Design: Floor # ______________________ Room # ___
Open ______________ Closed _____ Partitioned _____

(b) Floors:
Type ______________________
Covering ______________________
Condition ______________________
Odor ______________________
Observation ______________________
____

(c) Walls:
Type ______________________
Covering ______________________
Condition ______________________
Odor ______________________
Observation ______________________
____

(d) Ceiling:
Type ______________________
Covering ______________________
Condition ______________________
Odor ______________________
Observation ______________________
____

(e) Electrical Equipment:
Type ______________ Number ______________
Type ______________ Number ______________
Type ______________ Number ______________
Type ______________ Number ______________

(f) Air Supply:
Air Vents (# Type & Location) ______________________
____
____
Condition ______________________
Odor ______________________
Observations ______________________
____

Figure 12–1 (*Continued*)

Does occupant control air flow, temperature or humidity ?
How ______________________________
____

(g) Ergonomics :
Type of Lighting ______________________________
Glare Sources ______________________________
Temperature ______________________ Humidity ____
Do Windows Open ______________________________
Noise Source ______________________________
Vibration Source ______________________________
Room Dimensions: Sq. Ft. __________ Ceiling
Ht. __________
Work Operations ______________________________
____
Any Direct Chemical Use __________ What Type __________
____
Used For ______________________________
Do Any of the Occupants Smoke ______________________________
How Many ______________________________
Number of Room Occupants ______________________________
Any Plants in Room ______________________________

(h) Maintenance Schedule:
When Is the Room Cleaned ______________________________
With What ______________________________
By Whom ______________________________
Was the Room Recently Painted ______________________________
Any Other Maintenance ______________________________
Date __________ Type ______________________________

(i) Air Flow:
How Is the Room Air Distributed: H __________ V __________
Intakes and Exhaust Locations ______________________________
____
Any Cross Contamination ______________________________
Type ______________________________
Any Air Flow Blockages ______________________________
By What ______________________________
General Room Feeling (Stuffy, Clean, Musty, etc.) __________
____
____

Figure 12–1 (*Continued*)

Air Changes/Air Flow:
10 L/S __________ 17.5 L/S __________ 30 L/S __________
50 L/S __________ Other __________

Don't forget to check: ceilings, carpets, environmental surfaces, water reservoirs, microbiological sources, electrical equipment, sumps, false floors and ceilings.

Figure 12–1 (*Continued*)

This screening process will eliminate many possible sources for contamination of the structure's air supply and examine the building's physical stress aspects and design. It will also help target the more likely causes of poor indoor air quality for further investigation.

If the screening process authenticates the possibility of a building-related health problem, testing can then be implemented (See Figure 12–2). The building air quality and environment are analyzed predicated on the information determined in the initial investigation.

Testing can be an expensive proposition, depending on the building's size and depth of testing required. The elimination of potential problems and information developed in the investigative phase, on most likely causative factors, can significantly limit the testing necessary and reduce testing costs.

In theory, it is possible to cover all the bases and test for all likely contaminants and causes. Realistically, such practices are not practical or economic.

Testing will, however, involve two distinct areas: first, the building's air-quality handling characteristics including air-change rates, air-exhaust capacity, and air-flow characteristics; and, second, tests will be conducted to identify specific sources of contaminants.

Testing is conducted to clearly determine the root causes of the problem. When the causes are defined, remedial actions can be taken.

HOW IS A SICK BUILDING "CURED"?

This is definitely one case where a single "cure" will not apply to all situations. The "cure" will depend on what the causes of the

GRAB SAMPLE TESTS

(TEST FOR: CO_2, CO, NO_2, H_2S, SO_2, O_3, TCE, VOC, FORMALDEHYDE)

#	SAMPLE	LOCATION	RESULT

Use additional sheets as needed.

Figure 12–2 Air Quality Sample Log (Reprinted by permission of the authors, TCR, Inc. All rights reserved.)

problem are. Sick buildings are like people, individuals with their own unique sets of problems. Research has shown us that most sick buildings have some common traits, including the following:

- Generally the building is of an energy-tight design built in the 1970s with windows that don't open.
- There is poor air-handling equipment design, capacity, positioning, or maintenance.

- Current use is greater then originally designed; the work force has increased and there is additional electronic equipment.
- There is low ceiling height, poor office ergonomic design, and lots of low-frequency, strip-type lighting.
- Building interior design has been changed from its original plan.
- Occupants have minimal control over temperature, humidity, air flow, and lighting levels.

When the causes of a sick building are poor maintenance of the HVAC system or the presence of a chemical or biological agent, the "cures" are a straightforward attempt to eliminate the cause—to clean the office environmental surfaces and/or bring the air-handling systems up to snuff.

When the causes are external to the building or due to poor building design the "cure" is compounded. Replacement of air-handling systems, redesign of the structure, and elimination of external contaminant sources can be a long-term and expensive process.

In all cases, part of the cure will lie in education and information efforts. These efforts include:

- The effective handling of occupant complaints.
- Providing factual information to employees to allay fears.
- The developing of specific procedures and policies to audit buildings, identify problems areas, and address corrective actions.

Keep in mind that the "cure" also has its positive side. It brings the employer an opportunity to improve employee productivity, decrease sick time and employee turnover, and save money in the long run.

Ignoring the problem will also bring employers some things: losses, liability, and litigation.

The growing number of sick building syndrome-related litigations against building owners is having a pass-along effect to the contractor. Building owners faced with costly repairs for indoor air-quality problems and loss of rental income are seeking recovery against the architect, prime contractor, subcontractors, and even material suppliers involved in the building's construction.

Contractors can take some easy and simple steps to protect themselves:

- Understand the causes and methods to reduce SBS-related problems.
- Review plans, materials, and work operations with SBS problems in mind.
- Make written recommendations to owners and architects related to SBS prevention.
- Maintain air-quality monitoring logs and specific sample test logs to document baseline conditions during construction.

Chapter 13

Safety Resources

SAFETY INFORMATION RESOURCES

The most important piece of knowledge a person in the safety field can ever learn, when faced with a difficult question, is how and where to go to seek out necessary answers.

It's nice to know that you don't have to face a hostile world completely alone. There is in fact a wealth of information, contacts, and resources you can use. All you need to know is where to find them.

The listings within this section provide a starting point for increasing your circle of professional contacts and resources. Each group has its own area of expertise and special agenda and concerns. Yet, they all deal with one common thread—safety.

They all have information you can use and, best of all, people you can talk to with unique expertise in their particular fields.

Most have catalogues or lists of publications they make available, programs they sell, and information they distribute on safety and health issues.

The listings in this chapter are by no means complete, but they are a good place to start. There are many additional organizations involved in occupational safety not listed here. There are also organizations outside the United States involved in safety.

There are numerous safety resources within your own community—safety and health professionals doing the same types of safety jobs as you, chapter organizations of national groups and associations, municipal state and federal agencies, schools, and a host of business supplying the safety needs of industry.

Safety resources are all around you; the key is to know where to find them and begin to use them.

U.S. GOVERNMENT AGENCIES

Centers for Disease Control
Atlanta, GA
404-639-3311

Dept. of Health and Human Services
200 Independence Avenue, S.W.
Washington, D.C. 20201
202-443-2403 or 202-245-6296

Department of the Interior
Bureau of Land Management
Denver Federal Center, Bldg. 50
Denver, CO 80225
303-236-6452

Department of the Interior
Bureau of Reclamation, Room 7654
Washington, D.C. 20240-0001
202-343-4157

Environmental Protection Agency
Waterside Mall
401 M Street, S.W.
Washington, D.C. 20460
202-382-2090

Government Printing Office
U.S. Superintendent of Documents
Washington, D.C.
202-783-3238
Local GPO bookstores exist in many cities

National Institute of Environmental Science
Raleigh, NC 27611
919-541-3345

National Institute for Occupational Safety and Health
1600 Clifton Road, N.E.
Atlanta, GA 30333
404-639-3061

U.S. Department of Commerce
Census of Construction Industries
Available from the U.S. Government Printing Office

EPA REGIONAL OFFICES

Region 1—Boston MA	Phone: 617-565-3715
Region 2—New York, NY	Phone: 212-264-2525
Region 3—Philadelphia, PA	Phone: 215-597-9800
Region 4—Atlanta, GA	Phone: 404-347-4727
Region 5—Chicago, IL	Phone: 312-353-2000
Region 6—Dallas, TX	Phone: 214-767-2600
Region 7—Kansas City, KA	Phone: 913-236-2800
Region 8—Denver, CO	Phone: 303-293-1603
Region 9—San Francisco, CA	Phone: 415-978-8071
Region 10—Seattle, WA	Phone: 206-442-5810

Occupational Safety and Health Administration

The list below gives the addresses for OSHA. Information on field office locations can be obtained from the regional office locations.

U.S. Department of Labor
Occupational Safety and Health
Administration
200 Constitution Avenue
Washington, D. C. 20210
202-523-8271 & 202-523-8151

The Washington office of OSHA can provide information of specific standards, Voluntary Protection Programs, Field Operation Manual Information, and can answer technical questions. OSHA has a technical library that will assist in reference questions and a number of free publications. OSHA also publishes Fatal Facts on accident fatalities and technical bulletins on OSHA developments and announcements.

OSHA REGIONAL OFFICES

Region 1—OSHA: Connecticut, Maine, Massachusetts, New Hampshire, Rhode Island, Vermont
1 Dock Square Building, 4th Floor
16–18 North Street
Boston, MA 02109
617-223-6710

Region 2—OSHA: New Jersey, New York, Puerto Rico, Virgin Islands
1515 Broadway, Room 3445
New York, NY 10036
212-944-3432

Region 3—OSHA: Delaware, District of Columbia, Maryland, Pennsylvania, Virginia, West Virginia
Gateway Building, Suite 2100
3535 Market Street
Philadelphia, PA 19104
215-596-1201

Region 4—OSHA: Alabama, Florida, Georgia, Kentucky, Mississippi, North Carolina, South Carolina, Tennessee
1375 Peachtree Street N.E., Suite 587
Atlanta, GA 30367
404-347-3573

Region 5—OSHA: Illinois, Indiana, Michigan, Minnesota, Ohio, Wisconsin
230 South Dearborn Street, Room 3244
Chicago, IL 60604
312-353-2220

Region 6—OSHA: Arkansas, Louisiana, New Mexico, Oklahoma, Texas
525 Griffin Square Bldg., Room 602
Dallas, TX 75202
214-767-4731

Region 7—OSHA: Iowa, Kansas, Missouri, Nebraska
911 Walnut Street, Room 602
Kansas City, MO 64106
816-374-5861

Region 8—OSHA: Colorado, Montana, North Dakota, South Dakota, Utah, Wyoming
Federal Building, Room 1554
1961 Stout Street
Denver, CO 80294
303-837-3061

Region 9—OSHA: Arizona, California, Hawaii, Nevada, American Samoa, Guam, Trust Territory of the Pacific Islands
450 Golden Gate Ave., Box 36017
San Francisco, CA 94102
415-556-7260

Region 10—OSHA: Alaska, Idaho, Oregon, Washington
Federal Office Bldg., Room 6003
909 First Avenue
Seattle, WA 98174
206-442-5930

OSHA Training Institute
1555 Times Drive
Des Plaines, IL 60018
312-297-4810

Occupational Safety and Health Review Commission
1625 K Street
Washington, D.C. 20006
202-634-7960

Atlanta Regional Office — Phone: 404-347-4197
Boston Regional Office — Phone: 617-223-9750
Dallas Regional Office — Phone: 214-767-5271
Denver Regional Office — Phone: 303-844-2281

U.S. Small Business Administration
Office of Advocacy
1725 I Street, N.W.
Suite 403
Washington, D.C. 20006
202-634-6115

ADDITIONAL GOVERNMENT SAFETY RESOURCES

Advisory Committee on Construction Safety and Health (ACOCSH)
Occupational Safety and Health Administration
200 Constitution Avenue, N.W.
Washington, D.C. 20010
202-523-8651

National Institute of Drug Abuse (NIDA)
5600 Fishers Lane
Rockville, MD 20857
1-800-843-4971
1-800-622-HELP

National Technical Information Service (NTIS)
U. S. Department of Commerce
5285 Port Royal Road
Springfield, VA 22161
703-487-4929

TRADE AND INDUSTRY ASSOCIATIONS

Allied Stone Industries
P.O. Box 22478
Knoxville, TN 37922
615-966-6655

American Association for Clinical Chemistry
1725 K Street, N.W.
Washington, D.C.
202-857-0717

American Building Contractors, Inc.
729 15th Street, N.W.
Washington, D.C.
202-637-8800

American Concrete Institute
Box 19150/ Redford Station
22400 West 7 Mile Road
Detroit, MI 48219
313-532-2600

American Concrete Paving Association
3800 N. Wilke Road, Suite # 490
Arlington Heights, IL 60004
312-394-5577

American Concrete Pipe Association
American Concrete Pressure and Pipe Association
8320 Old Courthouse Road
Vienna, VA 22182
703-821-1990

American Concrete Pumping Association
1034 Tennessee Street
Vallejo, CA 94590
707-553-1732

American Consulting Engineers Council
1015 15th Street, N.W. # 802
Washington, D.C. 20005
202-347-7474

American Council for Construction Education
901 Hudson Lane
Munroe, LA 70201
318-323-2413

American Institute of Architects
1735 New York Avenue, N.W.
Washington, D.C. 20006
202-626-7300

American Institute of Building Design
1412 19th Street
Sacramento, CA 95814
916-447-2422

American Institute of Constructors
20 S. Front Street
Columbus, OH 43215
614-464-0598

American Institute of Timber Construction
11818 S.E. Mill Plain Blvd., Suite # 415
Vancouver, WA 98614
206-254-9132

American Institute of Steel Construction
400 N. Michigan Avenue
Chicago, IL 60611-4185
312-670-2400

American Construction Inspector's Association
2275 West Lincoln Avenue # B
Anaheim, CA 92801
714-772-7590

American Subcontractor's Association
1004 Duke Street
Alexandria, VA 22314
703-684-3450

American Welding Society
2501 N.W. 7th Street
Miami, FL 33125
305-443-9353

Associated Construction Publications
16227 W. Ryerson
New Berlin, WI 53151
414-782-0960

Association of Federal Safety & Health Professionals
7549 Wilhelm Drive
Lanham, MD 20706-3737
301-552-2104

Associated General Contractors of America, Inc.
1957 E. Street, N.W.
Washington, D.C. 20006
202-393-2040

Associated Specialty Contractors
7315 Wisconsin Avenue
Bethesda, MD 20814
301-657-3110

Construction Industry & Manufacturers Association
111 E. Wisconsin Avenue
Milwaukee, WI 53202
414-272-0943

Construction Management Association of American
12355 Sunrise Valley Drive Suite # 640
Reston, VA 22091
703-391-1200

Construction Specifications Institute
601 Madison Street
Alexandria, VA 22314-1791
703-684-0300

Federal Construction Council
c/o National Academy of Sciences
201 Constitution Avenue, N.W.
Washington, D.C. 20418
202-334-3378

Human Factor Society
Box 1369
Santa Monica, CA 90406
213-394-1811

Industrial Safety Equipment Association
1901 N. Moore Street
Arlington, VA 22209
703-525-1695

Mason Contractors Association of America
17 West 601 - 14th Street
Oakbrook Terrace, IL 60181-3799

National Aggregates Association, *also* National Ready Mix Concrete Association
900 Spring Street
Silver Spring, MD 20910
301-587-1400

National Asbestos Council
1777 N.W. Expressway, Suite # 150
Atlanta, GA 30329
404-633-2622

National Asphalt Paving Association
6811 Kenalworth Avenue, Suite # 620
Riverdale, MD 20737-1333
301-797-4880

National Association of Demolition Contractors
4415 W. Harrison Street
Hillside, IL 60162
312-449-5959

National Association of Dredging Contractors
625 I Street, N.W., Suite # 321
Washington, D.C. 20006
202-223-4820

National Association of Home Builders
15th & M Street, N.W.
Washington, D.C. 20005
202-8225-0200

National Association of Women in Construction
327 S. Adams Street
Ft. Worth, TX 76104
817-877-5551

National Constructor's Association
1101 15th Street, N.W., Suite # 1000
Washington, D.C. 20005
202-466-8880

National Construction Employment Council
1101 15th Street, N.W., Suite # 1040
Washington, D.C. 20005
202-223-1510

National Erectors Association
1505 Lee Highway, Suite # 202
Arlington, VA 22209
703-524-3336

National Precast Concrete Association
625 E. 64th Street
Indianapolis, IN 46220
317-253-0486

National Utility Contractors Association
1235 Jefferson Davis Highway, Suite # 606
Arlington, VA 22202
703-486-2100

Painting & Decorating Contractors of America
3913 Old Lee Highway, Suite # 33B
Fairfax, VA 22030
703-359-0826

Safety Equipment Distribution Association
1111 E. Wacker Drive, Suite # 600
Chicago, IL 60601
312-644-6610

Scaffold Industry Association
14039 Sherman Way
Van Nuys, CA 91405-2259
818-782-2012

Scaffold Shoring & Forming Institute
1230 Keith Building
Cleveland, OH 44115-2180
216-241-7333

Sheet Metal Air Conditioning National Association
Box 70
Merrifield, VA 22116
703-790-9890

UNIONS

Building & Construction Trades
Department, AFL/CIO
815 16th Street
Washington, D.C. 20006
202-347-1461

International Brotherhood of Electrical
Workers
1125 15th Street, N.W.
Washington, D.C. 20005
202-728-6137

United Association of Journeyman and
Apprentices of the Pipe Fitting
Industry
901 Massachusetts Avenue, N.W.
Washington, D.C. 20001
202-628-5823

United Brotherhood of Carpenters and
Joiners of America
101 Constitution Avenue, N.W.
Washington, D.C. 20001
202-546-6206

PROFESSIONAL ASSOCIATIONS

American Conference of Governmental
Industrial Hygienists (ACGIH)
6500 Glenway Avenue, Bldg D-7
Cincinnati, OH 45211
513-661-7881

American Insurance Service Group, Inc.
85 John Street
New York, NY 10038
212-669-0400

American Society of Safety Engineers
(ASSE)
1800 East Oakton
Des Plaines, IL 60018-2187
312-692-4121

National Safety Management Society
3871 Peadmont Avenue
Oakland, CA 94611
415-653-4148

World Safety Organization (WSO)
305 East Market Street
P.O. Box 518
Warrensburg, MO 64093
816-747-3132

PRIVATE ORGANIZATIONS

Bureau of National Affairs
1231 25th Street, N.W.
Washington, D.C.
1-800-452-7773
Publishers of: *BNA Job Safety And Health, Policy and Practice Series, BNA Occupational Safety and Health Reporter*

Business Round Table
200 Park Avenue, suite 222
New York, NY 10017
212-682-6370

Center for Construction Excellence
West Virginia University
Dept. of Civil Engineering
Morgantown, WV 26506-6101
304-293-3192

Commerce Clearing House
425 West Peterson Avenue
Chicago, Illinois 60646
312-583-8500
Publishers of *Employment Safety and Health Guide*

Construction Industry Institute
Department of Civil Engineering
2340 GG Brown Building
Ann Arbor, MI 48109
313-764-8496

National Fire Protection Association
(NFPA)
Batterymarch Park
Quincy, MA
617-770-3000

National Safety Council
444 Michigan Avenue
Chicago, IL 66601
312-527-4800
1-800-621-7619

Index